Shopify Theme Customization with Liquid

Design state-of-the-art, dynamic Shopify eCommerce
websites using Liquid's powerful features

Ivan Djordjevic

BIRMINGHAM—MUMBAI

Shopify Theme Customization with Liquid

Group Product Manager: Aaron Lazar
Publishing Product Manager: Shweta Bairoliya
Senior Editor: Ruvika Rao
Content Development Editor: Rosal Colaco
Technical Editor: Karan Solanki
Copy Editor: Safis Editing
Project Coordinator: Deeksha Thakkar
Proofreader: Safis Editing
Indexer: Pratik Shirodkar
Production Designer: Ponraj Dhandapani

First published: September 2021

Production reference: 1220921

Published by Packt Publishing Ltd.
Livery Place
35 Livery Street
Birmingham
B3 2PB, UK.

ISBN 978-1-80181-396-9

www.packt.com

To my mother, Sladjana Djordjevic, and the memory of my father, Jugoslav Djordjevic. Thank you for guiding me and helping me become the person I am today.

– Ivan Djordjevic

Contributors

About the author

Ivan Djordjevic comes from the small town of Prokuplje, Serbia. As a self-taught developer, he spent the first few years of his career working on different projects, but only when he came into contact with Shopify and learned about Liquid did he find himself. In 2016, Ivan joined Shopify Experts under HeyCarson, where he moved to the lead developer position after a few months. Since joining the Shopify family, Ivan found his passion in sharing his knowledge with other developers and guiding them on their way to becoming Shopify experts.

> *I want to thank all the people who have supported me through the years, especially Milos Ristic, Aleksandra Cugalj, and Nikola Skobo. You have been an inspiration and I am proud to call you my friends!*

About the reviewers

Nikola Škobo is a developer and manager based in the beautiful city of Belgrade, Serbia. He has over 10 years of experience in development, and since 2016, he's been working exclusively on the Shopify platform as a product member of Carson. Starting as a front-end developer, he climbed his way up through various senior-level roles until becoming the head of operations. With this, he helped Carson to grow from a small-task service to a premium-level service, helping thousands of merchants in the process.

Nikunj Suthar is the founder and CEO of Sibyll Software Private Limited and also a CTO at Webinopoly Inc. Before becoming a company founder in July 2016, Nikunj was employed as a project manager in an IT company and was responsible for all of the company's operations, including end-to-end management of project's meetings, daily activities, and service and support in all markets and countries.

Prior to founding Sibyll Software, Nikunj was also the CEO and was responsible for managing all of the projects. Nikunj earned a diploma from the **Institution of Electronics and Telecommunication Engineers** (**IETE**). Nikunj is also passionate about writing poetry in Hindi and Urdu. He is a well-rounded individual who lives with passion, dedication, and grace.

Table of Contents

Preface

Section 1: Shopify Explained

1

Getting Started with Shopify

Technical requirements	4	Managing a theme	12
What is Shopify?	4	Understanding theme structure	20
How to start?	5	Header	20
Navigating the admin panel	8	Sidebar	22
Core aspects	8	Summary	29
Sales channel	10	Questions	29
Settings	12	Further reading	30

2

The Basic Flow of Liquid

Technical requirements	32	Boolean	39
What is Liquid?	32	Nil	40
Understanding Liquid and its delimiters	32	Array	42
		EmptyDrop	43
Learning the comparison operators	34	Controlling whitespace	48
Working with logic operators	36	Summary	49
Understand the types of data	38	Quiz	49
Strings	38		
Number	39		

Section 2: Exploring Liquid Core

3

Diving into Liquid Core with Tags

Technical requirements	54	The for/else tags	71
Getting things ready	54	jump statements	73
Creating the product page	54	The for parameters	74
Creating the collection page	57	The cycle tag	78
Updating the navigation menu	58	**Theme tags**	81
Controlling the flow of Liquid	58	The layout tag	82
The if/else/elsif tags	58	The liquid and echo tags	83
The and/or tags	62	The form tag	85
The case/when tags	63	The paginate tag	87
The unless tag	65	The render tag	88
		The raw tag	91
Variable tags	66	The comment tag	91
The assign tag	66		
The capture tag	67	**Deprecated tags**	92
The increment tag	68	**Summary**	92
The decrement tag	70	**Questions**	93
Iterations tags	71		

4

Diving into Liquid Core with Objects

Technical requirements	96	**Content and special objects**	124
Working with global objects	96	The content_for_header object	124
Custom collection	98	The content_for_index object	125
Custom navigation	105	The content_for_layout object	125
Product customization	111	**Summary**	126
Improving the workflow with metafields	117	**Questions**	127
		Practice makes perfect	127
Setting up a metafields app	118	Project 1	128
Rendering the metafields value	122	Project 2	129

5

Diving into Liquid Core with Filters

Technical requirements	132	Product discount price	166
Working with HTML and URL filters	132	Exploring the additional filters	170
Building a product gallery	135	The default filter	170
		The t (translation) filter	170
Enhancing the product media gallery	140	The JSON filter	172
Building product accordions	153	Summary	173
The split filter	155	Questions	173
The index filter	163	Practice makes perfect	174
		Project 3	174
Math and money filters	164		

Section 3: Behind the Scenes

6

Configuring the Theme Settings

Technical requirements	180	The paragraph type	218
Exploring JSON settings	180	Glancing at the deprecated settings	219
Learning about the input setting attributes	184	The font input	219
Basic input types	184	The snippet input	220
Specialized input settings	194		
		Summary	220
Organizing the theme editor	216	Questions	221
The header type	216		

7

Working with Static and Dynamic Sections

Technical requirements   224
Static versus dynamic sections   224
Working with the section
schema   228
The name attribute   228
The class attribute   230
The settings attribute   230
The presets attribute   232

Building with blocks   234
The max_blocks attribute   242

Enhancing pages with JSON
templates   243

Building a JSON template structure   244
Upgrading a JSON template with
metafields   252

Exploring section-specific tags   256
The stylesheet tag   256
The style tag   256
The javascript tag   259

Summary   260
Questions   260
Practice makes perfect   260
Project 4   261

8

Exploring the Shopify Ajax API

Technical requirements   264
Introduction to the Shopify Ajax
API   264
Updating the cart session with
a POST request   265
The /cart/add.js endpoint   266
The /cart/update.js endpoint   272
The /cart/change.js endpoint   273
The /cart/clear.js endpoint   276

Retrieving data with a GET
request   277
The /cart.js endpoint   278
The /products/{product-handle}.js
endpoint   281
The /recommendations/products.json
endpoint   281
The /search/suggest.json endpoint   286

Summary   288
Further reading   289

Assessments

Appendix

Other Books You May Enjoy

Index

Preface

Nowadays, we usually learn new material using the various information found on forums and blogs, and we rely on the information we come across. Depending on the time spent on those articles, you will often skip some fundamental concepts, leaving even experienced programmers surprised when we discover some new and easier way to do something. This book covers the basic knowledge of Liquid and some advanced concepts that will set you on a proper path toward adding Liquid to your portfolio.

Liquid is an open source project created by Shopify co-founder and CEO Tobias Lütke. As a template language, Liquid variables connect the Shopify store's data to the static HTML content in our theme, allowing us to turn a static template into a fully dynamic and powerful e-commerce store, producing impressive results. Since 2006, Liquid has been growing and evolving. Today, many different web applications rely on Liquid, Shopify being one of those, which shows that the need for Liquid is ever-growing.

We can divide the book into three major sections. In the first section, we will get familiar with some basic information about Shopify, understand the Shopify interface and its theme structure, and start familiarizing ourselves with Liquid. Even though these topics might not sound that important, they will allow us to understand Shopify and Liquid at the foundational level. By understanding these basics, we will learn how to face problems in these areas, which are inevitable and frequent.

The second section of the book we will dedicate entirely to Liquid's core features, without which we will not be able to create many wonderfully complex components. While we will not go over every object, tag, and filter that exists, we learn about the essence of Liquid core and draw attention to some commonly used development techniques.

The third and final section of the book will take us behind the scenes, where we will learn how to use JSON to create easily configurable options that are the soul of a Shopify e-commerce store. Finally, we will learn how to use the Shopify Ajax API combined with the Liquid features covered previously to create powerful features and make our code more dynamic.

Who this book is for

This book is for beginners and experienced CMS developers who want to learn about working with Shopify themes and customizing those themes using Liquid. Web developers working on designing professional e-commerce websites would also find this book useful. Besides familiarity with standard web technologies (HTML, CSS, and JavaScript), this book requires no prior knowledge of Shopify or Liquid. The book covers everything from Shopify fundamentals, the core of Liquid, and the REST API, all the way to the latest Liquid features that may be new even to proficient developers.

What this book covers

Chapter 1, Getting Started with Shopify, creates a solid foundation for understanding what Shopify is, how it all works, and other essential knowledge that, as practice shows, is often skipped. The approach of cutting the theoretical and jumping straight into the syntax might sound tempting, but as we all know, even the tiniest ripples can cause considerable problems in the future. While we will not go much into how Toby Lütke created Shopify, we will cover some essential topics, including creating a private development store, creating a child theme, explaining theme structure, and other essential topics that we need to know.

Chapter 2, The Basic Flow of Liquid, helps us learn what Liquid is and how to write it by explaining the Liquid syntax. We will also go over logical and comparison operators, what types of data we can use within Liquid, what handles are, and other essentials. By learning how to use logical operators and manipulate handle attributes, we will also learn how to combine them to create various dynamic features.

Chapter 3, Diving into Liquid Core with Tags, covers programming logic, or in short, tags. There are many types of tags, such as control flow tags, which allow us to output a block of Liquid code based on various conditionals through iteration tags that will repeatedly run a block of code. Additionally, we will learn about the different types of variable tags we can use to store data, as well as theme tags that will allow us to render theme-specific tags.

Chapter 4, Diving into Liquid Core with Objects, helps us understand what content objects are, why they are mandatory, and how to use them, which is the first step in creating future templates. After that, we will move to global objects, the knowledge of which will open a new door for us and output our dynamic content on any page. Lastly, we will learn what metafield objects are and utilize them to output unique content on any page.

Chapter 5, Diving into Liquid Core with Filters, is a crucial topic of Liquid core, which will allow us to create or manipulate different types of data, which is a compelling feature. With Liquid filters, we will gain a more profound knowledge of how Liquid works and how easily we can output dynamic data such as the image HTML tag, calculating the product discount, and handling font variants.

Chapter 6, Configuring the Theme Settings, helps us learn about JSON settings and why these files are so important to us. Later, we will move to a more direct approach to the topic by learning what types of settings we can use inside our JSON files and the difference between the basic and specialized input settings.

Chapter 7, Working with Static and Dynamic Sections, will help us understand what sections are and how to use them to create easily configurable content through the Shopify theme editor. We will learn how to create static and dynamic sections and their counterpart blocks, which play a considerable role in theme development. Lastly, we will learn more about the newly introduced JSON-type templates and how we can combine them with metafields to create genuinely unique and easily configurable content on any page.

Chapter 8, Exploring the Shopify Ajax API, takes us through the Shopify Ajax API and explains its requirements, limitations, and possible use cases. Additionally, we will learn how to make different types of requests to the Shopify API. We will learn how to retrieve product information, add any number of products to the cart session, and read the cart's current content. Lastly, we will learn how to make a request through the Shopify API and retrieve the necessary information to create the product recommendation and predictive search feature.

Appendix, Frequently Asked Questions, contains additional guidelines, advice, and answers to the questions common for developers who are just getting started on entering the world of Shopify and Liquid.

Assessments, contains the answers to the questions from all the chapters.

To get the most out of this book

Shopify is a hosted service, and it does not require any particular setup or software/hardware. However, to get the most out of the book, developers should already be familiar with basic HTML markup, CSS, and understand the JavaScript scripting language, which we will need later during this book.

If you are using the digital version of this book, we advise you to type the code yourself or access the code from the book's GitHub repository (a link is available in the next section). Doing so will help you avoid any potential errors related to the copying and pasting of code.

While it might be short, I wholeheartedly hope that you will enjoy this little adventure of ours and learn a few things along the way. My intention for the book was to offer a vast array of theoretical knowledge backed by real-life examples and suggestions on how to work with a real-life project, where you can see how it all ties up together.

Download the example code files

You can download the example code files for this book from GitHub at `https://github.com/PacktPublishing/Shopify-Theme-Customization-with-Liquid`. If there's an update to the code, it will be updated in the GitHub repository.

We have other code bundles from our rich catalog of books and videos available at `https://github.com/PacktPublishing/`. Check them out!

Code in Action

Code in Action Videos for this book can be viewed at `https://bit.ly/3nHIQtD`

Download the color images

We also provide a PDF file that has color images of the screenshots and diagrams used in this book. You can download it here: `https://static.packt-cdn.com/downloads/9781801813969_ColorImages.pdf`.

Conventions used

There are a number of text conventions used throughout this book.

`Code in text`: Indicates code words in text, database table names, folder names, filenames, file extensions, pathnames, dummy URLs, user input, and Twitter handles. Here is an example: "We can easily translate most of the attributes inside the `schema` tag by including the translation keys as the `name` attribute value."

A block of code is set as follows:

```
{% section "related-product-1" %}
{% section "related-product-2" %}
```

When we wish to draw your attention to a particular part of a code block, the relevant lines or items are set in bold:

```
{% schema %}
{
    "name": "Announcement bar"
}
{% endschema %}
```

Any command-line input or output is written as follows:

```
_9VUPq3SxOc
youtube
```

Bold: Indicates a new term, an important word, or words that you see onscreen. For instance, words in menus or dialog boxes appear in **bold**. Here is an example: "Any section added to the theme via the **Add Section** button will allow us to include different content for any occurrence, and we can repeat it any number of times."

> **Tips or important notes**
> Appear like this.

Get in touch

Feedback from our readers is always welcome.

General feedback: If you have questions about any aspect of this book, email us at customercare@packtpub.com and mention the book title in the subject of your message.

Errata: Although we have taken every care to ensure the accuracy of our content, mistakes do happen. If you have found a mistake in this book, we would be grateful if you would report this to us. Please visit www.packtpub.com/support/errata and fill in the form.

Piracy: If you come across any illegal copies of our works in any form on the internet, we would be grateful if you would provide us with the location address or website name. Please contact us at copyright@packt.com with a link to the material.

If you are interested in becoming an author: If there is a topic that you have expertise in and you are interested in either writing or contributing to a book, please visit authors.packtpub.com.

Share your thoughts

Once you've read *Shopify Theme Customization with Liquid*, we'd love to hear your thoughts! Scan the QR code below to go straight to the Amazon review page for this book and share your feedback.

https://packt.link/r/1-801-81396-5

Your review is important to us and the tech community and will help us make sure we're delivering excellent-quality content.

Section 1: Shopify Explained

In this section, we'll go through some theoretical knowledge around Shopify, theme structure, creating a development store, and working on a theme. After familiarizing ourselves with Shopify, we will start learning about Liquid, how its syntax works, and how we can use logical and comparison operators to manipulate the various data types that we have at our disposal.

This section comprises the following chapters:

- *Chapter 1, Getting Started with Shopify*
- *Chapter 2, The Basic Flow of Liquid*

1
Getting Started with Shopify

From the very dawn of the internet, people saw the convenience of having information available at their fingertips. Ever since, people have been working hard on creating various internet applications and comprehensive services that will make our lives easier, and with these came e-commerce stores. Consequently, Shopify was born.

The first chapter sets a solid foundation for understanding what Shopify is and how it all works which, as practice shows, we often skip. The approach of cutting the theory and jumping straight into the syntax might sound tempting. However, even the tiniest ripples can cause considerable problems in the long run. While we will not go much into how Toby Lutke created **Shopify**, in this chapter, we are going to cover the following main topics:

- What is Shopify?
- How to start?
- Navigating the admin panel
- Managing a theme
- Understanding theme structure

By the end of this chapter, we will not only have learned what Shopify is, but we will also learn how to create an account under the Shopify Partners Program, create a development store to practice on, learn to navigate Shopify admin, and create a child theme and understand its structure. With this knowledge, we will have a solid foundation of how all these essentials will allow us to embark further on our learning journey to customize a theme on Shopify.

Technical requirements

While each topic will be explained and presented with the accompanying graphics, we will need an internet connection to be able to follow the steps outlined in this chapter, considering that Shopify is a hosted service.

What is Shopify?

Whether a developer or through general use of the internet, chances are that the name *Shopify* has come up at least once, *but what is Shopify?* **Shopify** is a multinational e-commerce company based in Ottawa, Canada, that offers various comprehensive services to its clients. This subscription-based service provides everything from buying the domain to building and managing a future dream store with ease.

Over the years of its existence, Shopify has proven that it is not just another store builder or a tool to sell products. Instead, it has established itself as an e-commerce powerhouse that allows anyone to build the store and create a unique experience for their shoppers. Being a template-based store builder, Shopify offers various free or paid themes to enable store customization, enabling you to use the intuitive and straightforward theme editor without any development knowledge. However, if the end goal is to create an utterly unique store with various customization options at our fingertips, then we will require a developer with Liquid knowledge to customize the theme and create additional options inside the code editor.

How to start?

The first step toward learning and working on Shopify requires us to learn about the Shopify Partners Program. The Partners Program is a platform created by Shopify that assembles people from all over the world and offers them the ability to build new e-commerce stores for store owners, design themes, develop apps, and refer new clients to Shopify. One of the most enticing extensions of the platform is that it will allow us to create a development store to practice our Liquid knowledge.

Before creating our **Development store** and familiarizing ourselves with Shopify, we will first need to create an account within the Shopify Partners Program. Creating the account is a relatively straightforward process, which we can start on the following page: `https://www.shopify.com/partners`:

1. We can begin the process by clicking on the **Join now** button in the top-right corner of the header or by clicking on the **Log in** button to access an existing account. After filling in the basic information on the create account page and the subsequent **Account Information** page, we will get our first view of the Partners dashboard:

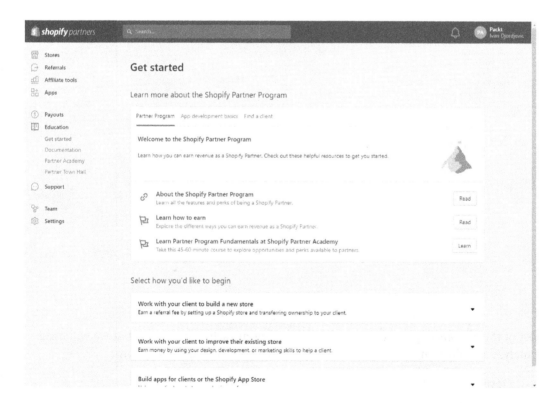

Figure 1.1 – Example of a Shopify partner dashboard

2.   After creating an account, the next step will be to create our **Development store**. We can do this by visiting the **Store** link in the upper-left sidebar and pressing the **Add store** button, which will prompt a store registration form that we will need to fill in:

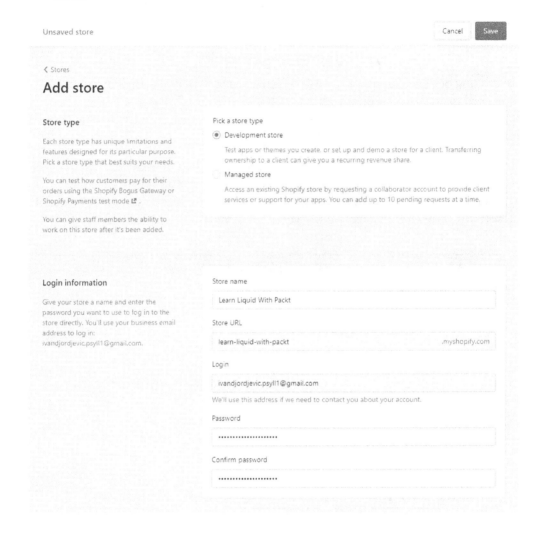

Figure 1.2 – Example of creating the Development store

Most of the options require no simplification as they are self-explanatory, but we will need to analyze the most important one, **Store type**, which consists of two options:

- **Development store**
- **Managed store**

For our learning purposes, we will be using **Development store**. However, we will also provide a short explanation of what the **Managed store** option means and when we can use it.

The **Managed store** option will enable us to request **collaborator** access to an existing store. Selecting this option will prompt another section named **Permission**, where we will need to choose the type of access that we are requesting. Generally, we can choose to request access to all areas of the client store. However, to perform theme customization on a client store, we will need access to the **Themes** option under **Online store**.

Once we have selected the permission level that we require to perform the work and have entered the URL of the store that we are requesting access for, all that is left is to send the request by pressing the **Save** button. The store owner will receive a notification of the store access request and then choose to grant or deny access.

The **Development store** options will allow us to create an entirely new store where we can practice our Liquid skills, test out new features before they are rolled out to the Shopify live storefront, or create a new store that we can later transfer to our client.

> **Important note:**
> While creating the **Development store** is completely free of charge, in transferring the store to the client, the theme will automatically lose development status. The person to whom we have transferred the ownership will need to select a recurring plan for which the developer who initially created the theme will receive a recurring commission for as long as they are paying for their subscription plan.

Let's create our development store by selecting **Development store** as the store type and filling in the store information.

> **Tip:**
> The developer preview is a new feature that Shopify has rolled out recently and will allow us to preview the innovations they are rolling out before they hit the live storefront. However, we will abstain from using this option.

Once you have filled in the login information, store address, and have selected the purpose of the store, press the **Save** button to create your store.

Success! Our store is ready to go, and with it, we are ready to proceed to familiarizing ourselves with the Shopify admin panel.

Navigating the admin panel

With the creation of a new development store, we will get the first glimpse of our store home page. The home page consists of the middle screen, which will contain some general advice for store owners to start their business, daily tasks, recent activity, and, on the left side of the screen, we will see the sidebar, which will be our primary focus.

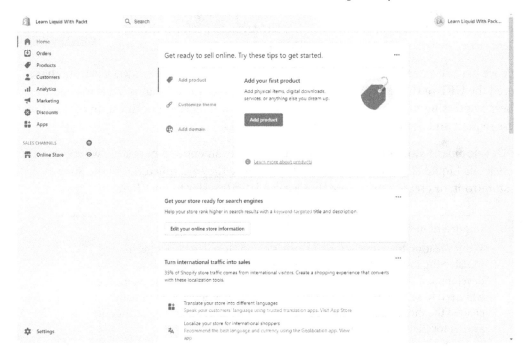

Figure 1.3 – Example Shopify admin home page

While we will not go into much detail straight away, it is imperative to have a basic understanding of each area of the store that we will be referencing later through different chapters of the book. We will split the sidebar into three sections for ease of reference, and we can list them as follows:

- Core aspects
- Sales channel
- Settings

Core aspects

The first section, **Core aspects**, contains the store-related options that the store owner generally uses, and this includes the following information:

- The **Orders** section contains all the information regarding the orders that the store owner has received. It is here that the store owner can preview each order individually and decide whether they would like to proceed with it. The owner may also manually create orders using the draft method and preview the abandoned checkout.

- The **Products** section consists of four separate areas that allow us to create and manage products, handle incoming inventory for the existing products, create and manage collections or product categories if you prefer, and create gift cards.

- The **Customers** section, as the name suggests, allows us to create and manage our customer database. This section, along with the previous **Products** section, will be of particular interest to us. We will return to them later with a detailed explanation of some of their functionalities that will be of interest to us.

- The following section, named **Analytics**, will primarily be used by the store owner. It offers a great deal of information regarding the store's performance, such as a detailed report on sales, along with a live view of the customers visiting the store and their behavior.

- The **Marketing** section, as its name suggests, allows us to view the store market strategy overview. We can create and manage campaigns via emails or other social networks and create automation to increase the store retention rate.

- We can use the **Discount** section to create a discount coupon code that we can share with our customers to manually enter the checkout to receive a discount on their complete order or a specific product. Additionally, we can make an automatic discount that will automatically trigger once we fulfill the requirements set by the store owner.

- The last section, named **Apps**, grants us a quick preview of all the apps installed on our store where we can manage or remove the apps if we choose to.

> **Important note:**
> Due to the Shopify platform's limitations, it is impossible to combine the discounts. If we have qualified for the automatic discount of 10%, we will not be able to enter the manual discount coupon code for free shipping that we have previously received from the store owner.

> **Tip:**
> While it is possible, you should never install an app on your own when working under a collaborator account on the managed type of store. Suppose you require a particular app to perform a task entrusted to you. In that case, you should reach out to the store owner, explain the need for it, and ask them to install it for you before proceeding, as they will need to grant specific permissions and share the store data with the app that should be reviewed and accepted by the store owner.

Sales channel

The second section of our sidebar, **Sales channel**, represents the various platforms we can use to sell our store products. By default, the only visible channel is **Online store**, which will be our primary objective; however, we can easily add more by clicking on the plus button next to the sales channel.

The **Online store** channel is the heart of the store as it provides us with the ability to output a visible storefront for our customers, and we can break it down into six individual sections:

- The **Themes** section allows us to manage our store look by customizing our store theme to our unique brand. The first thing that we can sometimes see once we open the **Themes** section is the note from Shopify that our online store is *password-protected*, meaning that the store is not yet visible to our customers. While password protection is in place, every visitor who tries to view our store will only see a notification that the store is password-protected and is currently inaccessible.

> **Important note:**
> We can easily disable password protection by clicking the **See store password** button on the password protection note in the **Themes** section or by visiting the **Preferences** section in the **Online Store** section. However, considering that we have selected the development option as our store, disabling password protection is unavailable. We can only remove password protection after transferring the store to our client or purchasing a subscription plan of our own.

After the password protection note, the next area that we can see under our **Themes** section is named **Current theme**. This section shows us the name and a small preview of the current theme, followed by the **Online store speed** section, which provides us with our store's speed report. The **Online store** speed section is currently disabled for password-protected stores.

Following the previous section, near the bottom of the screen, we can find the last section named **Theme library**. We can easily explore free and paid themes within this area by selecting their respective links or uploading our custom-made theme by choosing the **Upload theme** button.

- The **Blog posts** section allows us to manage and create blog posts that we would like to show on our store and categorize them under different blogs.

- The **Pages** section allows us to create multiple pages that our customers will be visiting frequently, such as the *About us* or *Contact us* pages, or the pages supporting our products by offering in-depth information. For more information on managing pages, refer to `https://help.shopify.com/en/manual/online-store/pages`.

- Inside the **Navigation** section, we can find the necessary tools that will allow us to create navigation with up to two-level nested menus that our customers can use to navigate our online store. For information on creating the navigation menu and managing the link list, refer to `https://help.shopify.com/en/manual/online-store/menus-and-links`.

- The **Domains** section shows us our current **primary domain**, which uses the format `my-store-name.myshopify.com`. Additionally, we can purchase a custom domain by using the **Buy new domain** button, or if we have obtained a domain through a third party, we can then set that as our primary domain. For more information on domains, refer to `https://help.shopify.com/en/manual/online-store/domains`.

- The last and final section under our **Online store** channel is named **Preferences**. As most of the options under this section are self-explanatory, we will not be going into too many details to keep the book to the point. However, if you would like to read more about each of them, you can visit their respective pages, which we will list, to get additional information. The **Preferences** section allows us to regulate some important settings that will help the store owners with their future store, and we can list them in the following way:

  - **Title and meta description**: For detailed information on the title and meta description, refer to `https://help.shopify.com/en/manual/online-store/setting-up/preferences#edit-the-title-and-meta-description`.

  - **Social sharing image**: For detailed information on social sharing images, refer to `https://help.shopify.com/manual/using-themes/change-the-layout/theme-settings/showing-social-media-thumbnail-images`.

- **Google Analytics**: For detailed information on Google Analytics, refer to `https://help.shopify.com/manual/reports-and-analytics/google-analytics`.

- **Facebook Pixel**: For detailed information on Facebook Pixel, refer to `https://help.shopify.com/manual/promoting-marketing/facebook-pixel`.

- **Customer privacy**: For detailed information on customer privacy, refer to `https://help.shopify.com/en/manual/your-account/privacy/cookies`.

- **Password protection**: For detailed information on password protection, refer to `https://help.shopify.com/manual/using-themes/password-page`.

- **Spam protection**: For detailed information on spam protection and usage of Google reCaptcha, refer to `https://help.shopify.com/en/manual/online-store/setting-up/preferences#protect-your-store-with-google-recaptcha`.

Settings

The third and final section of our sidebar, named **Settings**, contains many options to help store owners set up and run their store. Due to the significant number of sections and subsequent options inside **Settings**, we will not be covering all of them, but we will mention some of the options that will be of interest to us in some of the following chapters of the book.

Managing a theme

We have already mentioned this topic under the **Themes** section, but *what is a theme?* The **theme** is a master template file that controls your store layout, which allows us to change the storefront layout by editing the code or editing the theme editor settings through this template.

By default, a debut theme is automatically added as a starting theme when creating a new store. However, for our learning purposes, we will try and install a theme of our own:

1. The first thing that we need to do is to position ourselves in the **Themes** section inside the **Online store** sales channel. Once inside, scroll down until you reach the area named **Theme Library** and search for and click on the button called **Explore free themes**:

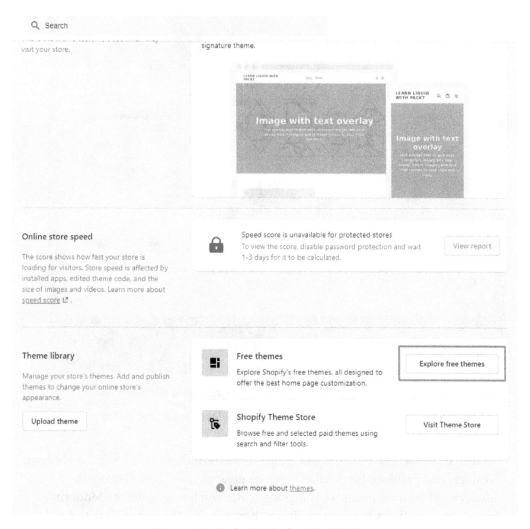

Figure 1.4 – Exploring the free Shopify theme

2. By pressing the **Explore free themes** button, we will see a popup containing the eight free Shopify themes to add to our store and the **Debut** theme, which we already have installed and is marked with the label **Current**:

Figure 1.5 – Selecting the free Shopify theme named Minimal

3. While we can choose any of these themes, let's select the second column theme inside the third row called **Minimal** by clicking on it.

    As with most themes, the **Minimal** theme contains multiple styles, such as **Modern**, **Vintage**, and **Fashion**. While all of these are great choices, we will select the **Modern** option for our learning purposes. Once you have chosen the **Modern** option, click on the **Add to theme library** button to finalize the process and add the newly selected theme to our store's theme library:

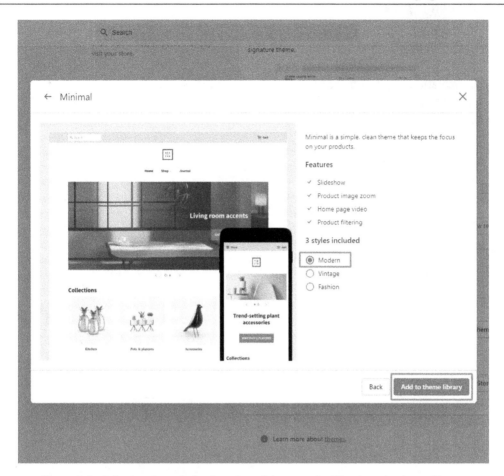

Figure 1.6 – Selecting the style for the selected theme and adding it to the theme library

Depending on your internet connection, it might take a few seconds while Shopify loads in your new theme, after which you will receive a notification that the **Minimal** theme was successfully added to your store. Even though we have added a new theme to our store, the **Debut** theme is still our live theme, whereas if we look under the theme library, we will see our newly added **Minimal** theme. To change that, we will need to set our new theme as the current theme.

4. We can publish a new theme live by scrolling down to the theme we are looking to publish live. In our case, that theme is **Minimal**. Click on the **Actions** button and then click on the **Publish** button, after which a popup will appear asking for confirmation to publish the **Minimal** theme live. Press the **Publish** button for the second time to confirm our choice, which will automatically publish and set the **Minimal** theme as our store's current theme:

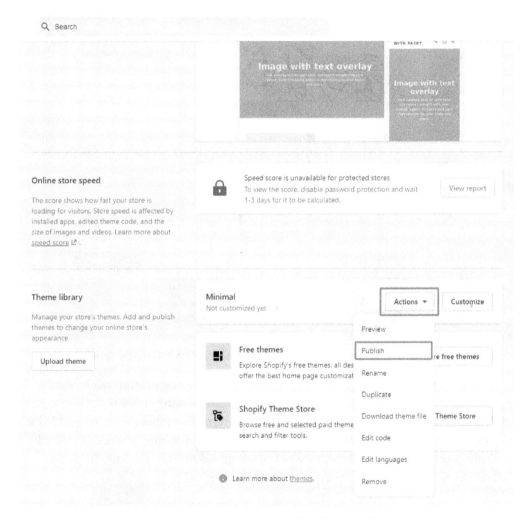

Figure 1.7 – Publishing a new Shopify theme live

5. Now that we have installed the new theme, it is time to preview how our new theme looks in our store. We can do this by clicking on the **Actions** button:

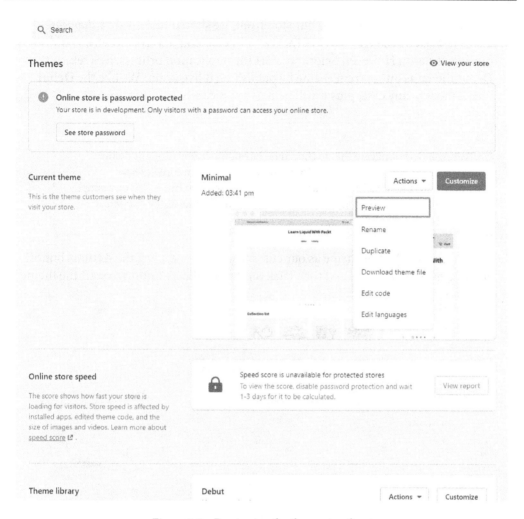

Figure 1.8 – Previewing the theme storefront

However, this time, the **Actions** button we should click on will be on our current theme, not our theme library, and then select the **Preview** button, which will open our store preview in a new tab.

As we can see from the preview page that has opened for us, the storefront does not look appealing, as it lacks content. We can only see some default sections with placeholder images.

6. Before we make any changes to our storefront, we should first create a *duplicate theme* to test our changes without the fear of it crashing our live storefront and causing us harm. However, before we start the duplication process, let's return the **Debut** theme as our current theme by publishing it live again. We like the **Debut** theme more in any case, plus it will help us practice what we just learned.

> **Tip:**
> Creating a new duplicate theme should be our number one thought before making any major modifications to our theme. Having multiple theme duplicates will help us pinpoint any potential issue caused by an app or a simple oversight that will break our live storefront, which will inevitably happen at some point.

7. After setting the **Debut** theme as our current store theme, click the **Actions** button on our current live theme and then click on the **Duplicate** button to start the theme duplication process:

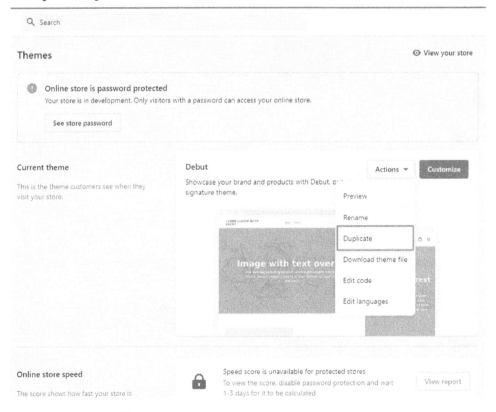

Figure 1.9 – Starting the theme duplication process

Considering that this is a newly installed theme without any content, the duplication process should not take too much time. After the process is complete, you will notice that we added a new theme to our **Theme library**, named `Copy of Debut`.

> **Important note:**
> Creating a duplicate theme each time you decide to make a significant modification is encouraged. However, we should keep in mind that Shopify only allows up to 20 duplicate themes per store. After we reach that limit, we will receive a notification that we have reached a limit of 20 duplicate themes per store. This limitation will also prevent us from creating a new theme duplicate, and if we want to make another one, we will need to delete some of the older theme duplicates that we no longer use.

8. By default, Shopify will automatically take the name of the theme we are duplicating and add the words `Copy of` in front of it. In view of the fact that having multiple similarly named themes can quickly get out of hand, we should immediately rename our new theme to avoid potential future confusion.

9. Click on the **Actions** button on our newly created **Copy of Debut** theme, and select the **Rename option**:

Figure 1.10 – Example of renaming the theme and its confirmation

This action will automatically prompt you with a popup where you can enter a new name. Shopify provides us with up to 50 characters to name our new theme, which gives us plenty of space to select a proper name. You should include supplementary information such as the date and what customization it contains. In our case, we will name the theme `Debut - Learning Shopify - 19 Apr '21`, and after, we will click on the **Rename** button to confirm our choice.

Now that we have learned how to create and rename a duplicate theme, it is time to dive into our newly created theme and learn more about how it works.

Understanding theme structure

To start familiarizing ourselves with the theme structure, we will first need to open the code editor. We can open the code editor by clicking on the **Actions** button on the Debut - Learning Shopify - 19 Apr '21 theme and then clicking on the **Edit Code** button, where we will have the first view of our code editor.

We can divide the code editor into the following two sections:

- Header
- Sidebar

Header

We can find the **Header** section at the top of the page, and it contains the name of the theme with the arrow button to exit the editor on the left side and three buttons on its right side, named as follows:

- **Preview**
- **Customize theme**
- **Expert theme help**

Preview

The first button on our list, the **Preview** button, will allow us to quickly preview the duplicate theme that we are working on and any changes that we have made. While the current preview of our theme contains only placeholder content, it does contain one element that will be of great use to us, the **preview bar**:

Figure 1.11 – Example of the preview screen and preview bar

We can find the preview bar at the bottom part of our preview screen, and it contains the name of the theme that we are previewing on the left side, and on the right side, it includes three buttons:

- The **close preview** button will automatically close our duplicate theme preview and redirect us to our live theme's home page.

- Clicking the **Share preview** button will trigger a popup that we can share with anyone to provide them with a glimpse of the changes we make on our same theme. While anyone who possesses this link will be able to view all aspects of your new theme, they will be unable to complete any purchases or reach the checkout page. In addition, the autogenerated preview link will only last for 14 days, meaning that after 14 days, you will need to generate a new preview link by repeating the preview steps and share it again with whomever you choose to.

- The last button in our preview bar is called **Hide bar**, and, as its name suggests, it allows us to hide the preview bar so that we can preview our changes without any visual obstacles. Note that the preview bar will automatically show itself when refreshing the page.

Customize theme

Under our editor's **Header** area, the next item is the **Customize theme** button, which will open another different kind of editor, a theme editor. Inside this editor, we can update some of the theme settings, such as typography, colors, and media links, and even manage the storefront sections.

Expert theme help

Finally, the **Expert theme help** button is something that the store owners will be using to post a job request using the Shopify Partners Program, whereby, at the end of this book, you will be in waiting as a Shopify expert.

Sidebar

The second section inside the code editor, **Sidebar**, lists all the files and directories we will be referencing; however, at the moment, we are unable to see all the directories. We can resolve this by clicking on the **Layout** and **Templates** directories to collapse them:

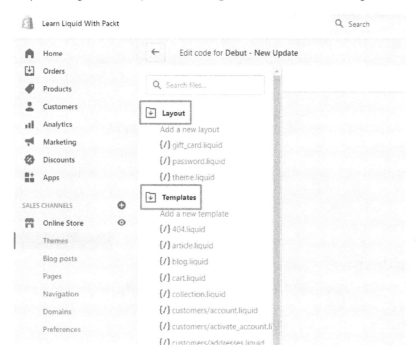

Figure 1.12 – Collapsing the directories inside the code editor

After collapsing the two directories, we will notice that the two arrows inside the directory icon are now gone and that we can now see the additional four directories that our theme contains. The Shopify theme contains the following directories:

- Layout
- Templates
- Sections
- Snippets
- Assets
- Config
- Locales

Layout

The Layout directory is the main directory of our theme as it contains the essential files that our store requires to work. This directory usually contains up to four files, which we can call theme layout templates, described as follows.

gift_card.liquid

gift_card.liquid is a template file containing the code that renders the gift card page and is later sent to our clients via email notifications when they purchase a gift card.

password.liquid

The password.liquid file template renders the online store password page that any of our customers will see if they visit our store while the store is in development mode. We mentioned what password protection is and how to disable it back in the *Navigating the admin panel* section when discussing **Sales channel** and its **Online store** area.

To better understand the password protection page, let's try and preview it to see how it works. You can preview your store password page by combining the URL of your store, https://my-store-name.myshopify.com, and adding the word /password at the end.

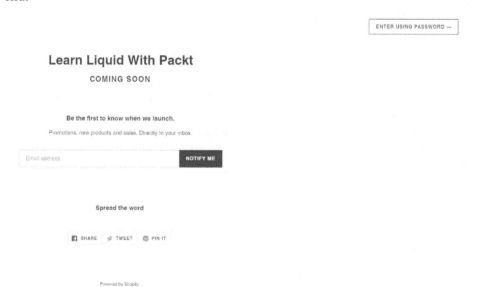

Figure 1.13 – Example of a Shopify password protection page

As we can see, password protection contains only the most basic information. However, it is successfully doing the work that we designed it for by preventing people from viewing our password, still in the Development store.

In the top-right corner of our password protection page, we will notice a button, **ENTER USING PASSWORD**, that will launch a popup where we can log in to our store using our store owner credentials and then click on the **Log in here** link, which will redirect us to our admin panel:

Figure 1.14 – Password page login form

However, what if we wanted to preview our theme storefront by entering the password? The password that this form is requesting is the same password that we have set to enable our password protection page. In our case, it was automatically generated by the system when we selected the development option as our store type.

To ascertain our password protection page's password, we need to return to our admin panel by typing https://my-store-name.myshopify.com/admin in the URL of our new browser tab. Once inside, in the left sidebar, under **Sales Channel**, expand the **Online store** field by clicking on it and then click on the **Preferences** option. Under the area named **Password protection**, you will find the password needed for our password page form to gain access to our storefront:

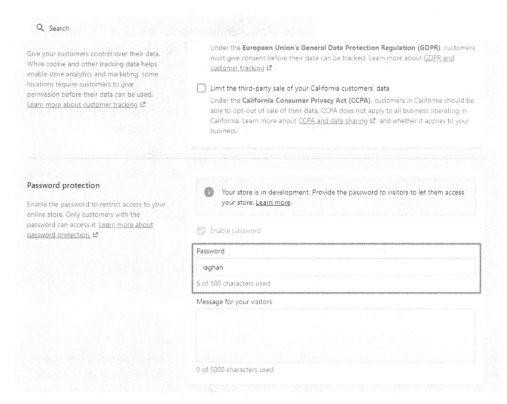

Figure 1.15 – Password protection page settings

Suppose we were to return to our password protection page and enter the password that we have found. In that case, the password protection page will be temporarily suspended for us, and the store will redirect us to the preview or live theme storefront, depending on the type of link we have initially opened.

theme.liquid

Back inside our code editor, we can consider our next item on the list, named theme. liquid, as the master layout file in which all other template files and any other element that we will learn of later will be rendered.

checkout.liquid

The last layout file on our list, named checkout.liquid, is not currently visible on our Development store. This layout file is only visible to the Shopify store owners who have purchased the **Shopify Plus** subscription.

Usually, each theme contains a set of predefined options that will allow us to make some basic styling changes to our checkout page. However, with the `checkout.liquid` layout file in our possession, we will have access to our checkout file, where we can create some more complex modifications that we would otherwise not be able to do.

Note that even with the `checkout.liquid` file in our hands, we will not be able to modify the flow of the checkout page process due to security reasons. We will only be able to make some basic modifications that will not interrupt the checkout flow.

We can only activate the Shopify plan by submitting a request to Shopify support. After reviewing our application, they will generate a custom price for enabling this unique plan in our store.

> **Important note:**
> Even though you have activated the Shopify Plus plan on your store, the `checkout.liquid` file will not be visible immediately. Instead, you will need to submit a request to Shopify support and ask them to include this unique file in your store.

For these reasons, we will not be going into too much detail regarding the `checkout.liquid` file. However, we will cover the *how-to* and most essential elements that the layout files contain, which should set us on a proper path of understanding layout files. For more information on editing the checkout file, refer to: `https://shopify.dev/themes/architecture/layouts/checkout-liquid`

Templates

The next point on our theme directories list is `Templates`, a group of files that allows us to create and manage the look of multiple pages all at once. `Templates` files consist of two types of files:

- The first type of `Templates` file is a `.json` type of file, which is a new addition to Shopify. Using the `.json` type of template, we can easily control the layout of any page through the theme editor. However, for better understanding, we will not go into too many details right now. It will be far more productive to cover the `.json` template and its possibilities in some of the following chapters.

- The second type is a `.liquid` type of file, which is a simple markup type of file with which we will familiarize ourselves right now.

Practice dictates that each theme comes with one template file for each page type, for example, product.liquid, which the system will automatically assign to any current or future product page that we may create. Considering that Shopify is a template-based file, any change made to a specific template will affect any page to which we have previously assigned this template. However, Shopify also allows us to create additional template files for each page type and customize them further without changing our original template file layout.

We can create a new template file by clicking on the **Add a new template** button located just below the template's directory, after which a popup will appear, asking us to choose a type and name for our new template file:

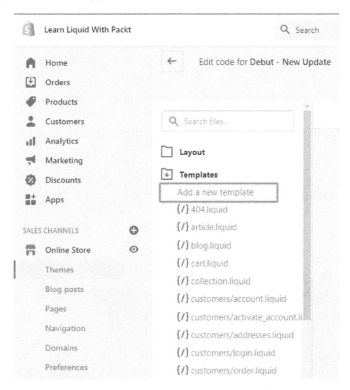

Figure 1.16 – Creating a new template file

After successfully creating a new template file, we can now assign the new template to the page for which we have created the new template. We can do this by opening any page inside our admin panel, depending on the type of page template we have created, and selecting the new template name inside the **Template suffix** drop-down menu located under the **Theme templates** area.

> **Important note:**
>
> The **Template suffix** drop-down menu can only read values from the current live theme. What this means is that the newly created template file will not be visible in our admin until we either publish our duplicate theme live or create the same template file within our current live team. If we opt for the second choice, note that we need to create the file with the same name; we do not have to make any changes to the file's content.

If we do not see the **Theme templates** area in our admin, we should confirm that we are on the right page by checking which template type we have created, as the **Template suffix** drop-down menu will only be visible on the pages that have more than one template created.

Sections and snippets

The next point of interest within our list of directories is called `Sections`, a different type of template file that, when combined with template files, allows us to create the genuinely customizable features for which Shopify has become famous. Note that any variable created inside the section will not be accessible outside the section, and vice versa. The only exception to this rule is that it offers a one-way communication if a section includes snippets.

The `Snippets` files allow us to re-use repetitive pieces of code over `Templates/Sections` by referencing their names. Besides allowing us to re-use parts of code, `Snippets` will enable us to access the variables inside the parent element as long as we pass those variables to the snippet as a parameter.

Assets

As the name suggests, the `Assets` directory allows us to store any theme-related assets, including images, font files, JavaScript, CSS files, and references them easily throughout the theme files.

Config

The `Config` directory is vital within our theme. Within this directory, we can define and manage the global JSON values for our theme. The directory consists of two key files:

- The `settings_schema.json` file allows us to create and manage the content inside the theme editor on our theme, which we can reference throughout the entire theme file.

- The `settings_data.json` file, on the other hand, records all the options defined in our schema file and saves their values. You can consider this file as your theme database, which will allow us to preview the current JSON values or modify them by updating the theme settings under the theme editor, or by directly editing the values inside the `settings_data.json` file.

Locales

The last directory on our list, named `Locales`, contains the theme locale file, which we can use to translate the content of our theme. The number of files that this directory may contain can vary. It can have one default file, `en.default.json`, or it can include multiple files depending on how many languages you would like to offer on your store.

Summary

In this first chapter, we have covered the essential aspects of Shopify by learning what Shopify is and how to create an account on the Shopify Partners Program, which we will use through our learning process, as well as for any future work on the Shopify platform.

We have created a development store and have also understood how and why it is crucial to create a duplicate theme before making any significant changes. While some of the things we have covered might sound irrelevant, each of these will help us to better understand the workflow that we will be doing regularly as a Shopify developer.

Lastly, we have acquired some knowledge of the internal structure of our theme files, which will be of great use to us in the following chapter, where we will be familiarizing ourselves with the fundamentals of Liquid and further developing our knowledge on theme customization.

Questions

1. What is the Partners Program?
2. How can we disable the password protection of the Development type store?
3. What is the difference between the `Layout` and `Templates` directory files?
4. Under what circumstances will the new template file be visible inside the admin section of your page?
5. What types of files and what conditions will allow us to access the variables within the parent file scope?

Further reading

- For additional information and an overview of Shopify admin, refer to `https://help.shopify.com/en/manual/shopify-admin`.

- For detailed information on managing your store account, refer to `https://help.shopify.com/en/manual/your-account`.

- For additional information on the online store, refer to `https://help.shopify.com/en/manual/online-store`.

2
The Basic Flow of Liquid

This chapter will help us understand what Liquid is and learn the basics of Liquid to provide us with the essential knowledge needed to master it. We will split the chapter into the following topics:

- What is Liquid?
- Understanding Liquid and its delimiters
- Learning the comparison operators
- Working with logic operators
- Understanding the types of data
- Controlling whitespace

By the end of this chapter, we will gain a deeper understanding of Liquid logic, the type of operators we can use to manipulate the various types of data, and the Liquid way of removing those pesky whitespace characters. By learning how to use logical operators and manipulate handle attributes, we will gain valuable knowledge of producing various dynamic features, setting our skills on the correct path toward writing quality and complex code.

Technical requirements

The code for this chapter is available on GitHub here: `https://github.com/PacktPublishing/Shopify-Theme-Customization-with-Liquid/tree/main/Chapter02`.

The Code in Action video for the chapter can be found here: `https://bit.ly/3ArKxia`

What is Liquid?

In the previous chapter, we gained the first insights into Shopify. We learned what Shopify is, how to create Shopify Partner accounts, and how to manage our theme. Finally, we have learned about theme structure, the directories it contains, and some essential files in our `Layout` directory, at which point we might have noticed that most of our files contain the `.liquid` extension. *So what exactly is Liquid?*

Liquid is an open source project created by Shopify co-founder and CEO Tobias Lütke. As a template language, Liquid variables connect the Shopify store's data to the static HTML content in our theme, allowing us to turn the static template page into a fully dynamic and powerful e-commerce store and producing impressive results. We can also consider Liquid elements as placeholders that will only get populated with proper data after the code inside the file is compiled and sent to the browser.

Ever since 2006, Liquid has been growing and evolving. Today, many different web applications rely on Liquid, Shopify being one of those, which shows that the need for Liquid is ever-growing. Learning Liquid is a great way to advance your knowledge further, and with its easy-to-learn syntax, we will master Liquid and be able to create complex functionality in no time.

Understanding Liquid and its delimiters

One of the two ways that we can discern a Liquid file is by the extension `.liquid`. Being a template language, a Liquid file is a combination of *static and dynamic content*:

- Elements that we write in HTML are called **static content**, and they stay the same no matter what page we are currently on.

- On the other side, elements written in Liquid are called **dynamic content** elements, whose content changes depending on the page we are on.

While our browsers can quickly process the HTML code, they would not know what to do with Liquid code as they do not understand it. We can break up the flow of what happens when we submit a Shopify URL to our browser into five logical steps:

1. The Shopify server tries to determine which store we are trying to access.

2. Depending on the type of page we are currently requesting information for, Shopify tries to locate and select the proper Liquid template from the active theme directory.

3. After successfully identifying the Liquid template that we require, the Shopify server starts replacing the placeholders with actual data stored on the Shopify platform.

4. Once Shopify has finished replacing the placeholders and performing any logic located inside the selected template, our browser will receive the compiled HTML file.

5. Now that our browser has received the HTML file as a response, the browser starts processing the file and fetches all other required assets, including JavaScript, stylesheets, images, and so on.

The second way that we can distinguish Liquid files and code is by its two delimiters:

- {{ }} double curly braces are used to indicate that we are expecting an output. An example Liquid code where we are expecting an output is as follows:

```
Our collection name is {{ collection.title }}
```

In the previous line, we see a string, followed by {{ collection.title }}. As we can see, collection.title is encapsulated inside the double curly braces indicating that the result of the code will be output. After the Shopify server processes our Liquid code and returns us something that our browser can work with, we would receive the following string as a result:

```
Our collection name is Winter Shoes
```

- {% %} curly braces with a percentage, on the other hand, are used if we want to indicate that we are looking to perform some kind of logic.

In our last example, we were able to see the result of using {{ collection.title }} to recover the name of the collection. Now, *what if we wanted to show the collection description, but for some reason, the* collection description *field did not return anything?* We would have ended up with an incomplete message:

```
Our collection description is
```

To ensure that this does not happen, we can check whether the data value exists using Liquid logic and comparison operators combined with some Shopify data types.

Learning the comparison operators

With Liquid, we have access to seven comparison operators, which we can combine to create any type of logical flow that our code requires. Let's review them as follows:

- The == operator allows us to check whether the element we are comparing is equal to some value:

```
{% if collection.title == "Winter Shoes" %}
  Winter Shows collection found!
{% endif %}
```

If collection.title is strictly equal to our string of "Winter Shows", the logic returns true, and we will see our message shown. Otherwise, the logic returns false. Note that comparison will only return true if the string is an exact match, including the text case.

- The != operator works similarly to the previous operator. The difference is that this operator checks whether the element we are comparing is not equal to some value:

```
{% if collection.title != "Winter Shows" %}
  Winter Shows collection is not found!
{% endif %}
```

As with the previous example, we will be using collection.title, but in this case, we will be checking whether the collection's name is not equal to the "Winter Shows" string. If the result is that collection.title is not the same as our string, the logic returns true, and we will see our message shown. Otherwise, the logic returns false, and the message will not be visible.

- The > operator allows us to check whether the compared value is greater than the comparing value:

```
{% if collection.all_products_count > 25 %}
  The collection has more than 25 products!
{% endif %}
```

Here, we are checking whether the number of all products in a collection is greater than 25. If it is, the logic will return true, and we will see our message shown. Otherwise, the logic returns false, and the message will not be visible.

- Similarly, as in the previous example, we will be checking the number of products in our collection, however, in this case, we will be using the < operator, which returns `true` only if the comparing value is less than the value being compared to:

```
{% if collection.all_products_count < 25 %}
  The collection has less than 25 products!
{% endif %}
```

If the number of products in the collection is less than 25, the logic will return `true`, and we will see our message shown. Otherwise, the logic returns `false`, and the message will not be visible.

We have a general understanding of how the < and > operators works, but what if we had the following example:

```
{% if collection.all_products_count > "25" %}
  The collection has more than 25 products!
{% endif %}
```

As we recall, the > operator allows us to check whether the compared value is greater than the comparing value. If it is, the logic returns `true`, and our message will show. However, if we were to run our code now, the logic would return `false`, and we would not see our message. *Why?*

If we look at our example, we will notice that we have quotation marks encapsulating our comparing value, which means that our comparing-to value is a string, compared to our `collection.all_products_count`, which returns a number. As we have mentioned before, the comparison operators will return `true` only when we have met the exact conditions that the operator requires. In our current example, for our message to be made visible, two conditions must be met.

The first condition is that both values must be the same data type, meaning that we cannot mix two types of data as we did just now. We would need to remove the quotation marks from our comparing-to value, consequently turning it from a string to a number data type.

The second condition is that our comparing value is greater than the compared to value. Once we have met the two conditions, the logic returns `true`, and our message will be shown.

- The >= operator allows us to check whether the element we are comparing is greater than or equal to the comparing value:

```
{% if collection.all_products_count >= 25 %}
  The collection has more or an exact number of
    products as the comparing value!
{% endif %}
```

Since both values are the same data type, if the number of products is greater or equal, the logic will return `true`, and we will see our message shown. Otherwise, the logic returns `false`, and the message will not be visible.

- The <= operator allows us to check if the compared value is less than or equal to the comparing value:

```
{% if collection.all_products_count <= 25 %}
  The collection has less or an exact number of
    products as the comparing value!
{% endif %}
```

Similarly, as with our previous comparing element, we check whether the total number of products in our collection is less than or equal to the compared value. If we have met the condition, the logic returns `true`, and our message will show. Otherwise, the logic will return `false`, and the message will not be visible.

Working with logic operators

Besides comparison operators, we also have access to two logic operators, which allow us to combine multiple conditions to create complex statements. We can divide them into the two following groups:

- The `or` operator allows us to set multiple conditionals, where we must meet at least one of them:

```
{% if collection.title == "Winter Shoes" or
  collection.all_products_count > 25 %}
  The collection name is Winter Shows, or the
    collection contains more than 25 products!
{% endif %}
```

In the preceding example, we check whether the name of our collection is equal to `Winter Shoes` or if the collection contains more than 25 products. If we have met at least one of these two conditions, the logic will return `true`, and our message will be shown. Otherwise, the logic will return `false`, and the message will not be visible.

- Similarly, the `and` operator allows us to set multiple conditions. However, for this operator to return `true`, all conditions must be met:

```
{% if collection.title == "Winter Shoes" and
  collection.all_products_count > 25 %}
  The collection name is Winter Shoes, and the
    collection contains more than 25 products!
{% endif %}
```

In this example, we are trying to see if both conditions have been met by checking whether the collection name is equal to `Winter Shoes` and the collection contains more than 25 products. If we have met both of these conditions, the logic will return `true`, and the message will be shown. Otherwise, the logic will return `false`, and the message will not be visible.

> **Important note:**
> We can now create a conditional with multiple comparison values. However, we should note the order in which our conditional will perform the check is from the right side. We cannot change this order using the parentheses, as parentheses are invalid characters in Liquid and will break our code.

- `contains` is the final operator on our list. This particular operator works differently than the previous ones as it does not check whether a comparing value is equal to a compared-to value. Instead, it allows us to check whether a string contains a substring:

```
{% if collection.title contains "Christmas" %}
  Our Christmas collection is stocked and ready to go!
{% endif %}
```

In the previous block of code, we check whether `collection.title`, which returns a string, contains the word `"Christmas"`, and if it does, it shows the message. Note that `contains` is case-sensitive and will only return `true` if the sub-string strictly matches the part of the string.

> **Important note:**
>
> We can use `contains` to search for the presence of a substring inside a string or even to check whether our string is a part of the array of strings. However, we cannot use it to check for the presence of an object inside the array of objects. We can only use `contains` to search for the presence of strings.

We have learned a lot about using logical and comparison operators. However, we only learned how to compare one value against another known value. *So how would we check if a compared value, for example a collection description, exists?* Since `collection.description` returns content with which we are unfamiliar, we have no way of performing this check with our current knowledge. To perform this action, we would first need to learn the different types of data we can use with Liquid.

Understand the types of data

So far, we have mentioned two data types, **strings** and **numbers** to be precise. However, in Liquid, there are six different types of data that are available to us:

- Strings
- Number
- Boolean
- Nil
- Array
- EmptyDrop

Strings

A **string** is a type of data that we use to represent text. Since a string can be any combination of letters, numbers, or special characters, we should always encapsulate it with quotation marks:

```
{% if product.title contains "Book" or product.title
  contains "2021" %}
  We have found a product that contains the word Book or
    the product that contains the word 2021.
{% endif %}
```

In the previous example, we check whether the product title contains the string `"Book"` or the same product title contains the string `"2021"`, and if it does, our message will be shown.

Number

A **number** is a type of data that require no quotation marks, and we use it to represent the following two types of numerical data:

- A **float** is a floating-point number, meaning that the number contains a decimal point.

- An **integer** or **int**, on the other hand, is a whole number without a decimal point:

```
{% if product.price > 25 and product.price < 3500 %}
    The number of products in a collection is greater
      than 25, and the product's price is lower than 3500.
{% endif %}
```

In this example, we check whether the product price is greater than 25, and at the same time, lower than 3500. If both conditionals are true, the message will be shown. Remember that since we used the and comparator, both conditionals must be true. If either one of them returns false, our message will not be shown.

Notice that neither of our two comparing values, 25 or 3500, are encapsulated inside the quotation marks. Otherwise, they would become strings, which, as we remember, we cannot compare against the whole number that product.price will return.

Boolean

Boolean is a type of data that can either be true or false:

```
{% if customer.accepts_marketing == true %}
  The customer has signed up for our newsletter!
{% endif %}
```

As we can see from the example, similar to the number data type, Booleans do not use quotation marks. If customer.accepts_marketing has signed up for our newsletter, the object will be true, and our message will be visible.

Nil

Nil is a data type that returns an empty value when our code has no results. Since nil does not return any values, we can use it to check for the truthfulness of a statement. However, we should note that we can only use nil to check whether the value exists. We cannot use nil to check the content of the value:

```
{% if customer %}
  Welcome {{ customer.name }}!
{% endif %}
```

In the previous example, we check whether the customer visiting our store has an account on our store using the conditional to perform the check if the customer object exists. Note that we have not added any operator or data type after our customer object. The reason for this is that nil is a special empty value. It is not a string with text `nil` or a variable and, as such, does not require any visual representation.

We can split the values inside Liquid into two categories:

- All data types that are true by default when used inside a conditional are considered **truthy**. Note that even empty values, such as strings, are considered truthy by default.

- On the other hand, all data types that return `false` are considered **falsy**. The only two values that return `false` by default are false and nil.

Since nil is considered falsy, our message will not show unless the customer has an account in our store. Otherwise, the conditional will return `true`, and our message will show.

While nil is extremely useful, it does not provide an answer to all of our questions. If we recall, in one of our previous examples, we used `collection.description` to see the content of our description collection:

```
Our collection description is {{ collection.description }}.
```

If our collection description has some defined content, our message will show correctly. However, if we have not defined a description for our collection, then `collection.description` will return empty, and we would end up with an incomplete message:

```
Out collection description is.
```

As we recall, we can only use nil to check whether an element such as an object exists. It cannot check its content, *but what exactly does that mean?*

```
{% if collection.description %}
  <h3>{{ collection.description }}</h3>
{% endif %}
```

In this example, we have wrapped our element within a conditional using the nil data type to check whether a `collection.description` instance exists. However, as we can see, this is not the result that we hoped for:

```
<h3></h3>
```

Since nil only checks whether an element exists, in our case, our conditional has found that `collection.description`, while empty, does exist. Remember that all values inside Liquid except false and nil are by default truthy, meaning that even empty strings are considered truthy. For this reason, even though our `collection.description` instance was empty and returned an empty string (since we used nil inside a conditional), the result was that the value was truthy. Therefore, the code inside the conditional is visible.

To resolve this problem, instead of checking whether the value exists, we should check whether our value is not blank:

```
{% if collection.description != blank %}
  <h3>The collection description element will show only if
    its content is not empty.<h3>
{% endif %}
```

In the previous example, we check to see whether our collection description is not empty using the not equal parameter against `blank`. If the result of this conditional is that `collection.description` is not blank, the message we have defined will show. Otherwise, the conditional will return `false` and the message defined inside the conditional will not render.

Array

An **array** is a data structure that contains a list of elements, usually of the same type. Note that we cannot initialize our array using Liquid only, but we can break down a string into an array of substrings, the data in which we can access using one of these two methods:

- The first option for accessing data inside an array is by directly accessing each item individually.

  Since, as we mentioned, we cannot initialize an array using Liquid only, for this example, we can use `product.tags`, which would return an array of strings:

  ```
  "learning", "with", "packt", "is", "awesome!"
  ```

  To access these strings inside this string, we can use a square bracket annotation `[]` combined with `product.tags` and the index position to access each item individually:

  ```
  <p>{{ product.tags[0] }}</p>
  <p>{{ product.tags[1] }}</p>
  <p>{{ product.tags[2] }}</p>
  <p>{{ product.tags[3] }}</p>
  <p>{{ product.tags[4] }}</p>
  ```

  Note that array indexing starts at 0, so we use `product.tags[0]` to access the first element in our array. After submitting our code to the Liquid server, we receive the following list of strings:

  ```
  <p>learning</p>
  <p>with</p>
  <p>packt</p>
  <p>is</p>
  <p>awesome!</p>
  ```

  While this method produces results, it is only applicable when we are entirely aware of the content of our array and the exact position of the element that we require. If, on the other hand, we are looking to output the entire content of an array without too much writing, we would need to use a different method.

- To access all items inside an array, we would need to loop through each item in our array and output their contents:

  ```
  {% for tag in product.tags %}
      <p>{{ tag }}</p>
  {% endfor %}
  ```

We can use the `for` tag to repeatedly execute a block of code or to go over all of the values inside an array and output them:

```
<p>learning</p>
<p>with</p>
<p>packt</p>
<p>is</p>
<p>awesome!</p>
```

As we can see, the results of running our `for` loop are precisely the same as when we previously accessed each item individually.

While it is easier to use the loop over an array to quickly output an array than to call out each item inside an array individually, it is essential to be aware of both methods, which will allow us to create some complex functionalities later on.

EmptyDrop

The final data type in our list is **EmptyDrop**, which would result from us trying to access an attribute of the previously deleted object or disabled by using its handle.

Before we learn about the final data type, we will first need to learn what a handle is, and how to find it.

Finding the handle

The **handle** is a page title written in lowercase, where hyphens (-) are used to replace any special characters and spaces. Similar to when we learned how to access an individual item inside an array using the position index, we can use page handles to access the attributes of a Liquid object.

We now understand what a page handle is, but let's try to put that into context by creating a page on our store and see what we just learned in action:

1. To create a new page, we will first need to navigate to our store by visiting `https://www.my-store-name.myshopify.com/admin` and logging in using our Shopify Partner credentials. An alternative way to do this is to visit the Shopify Partners site at `https://www.shopify.com/partners` and log in with the Partner account that we created in the previous chapter, subsequently clicking on the **Store** button in the left sidebar and then logging in to our store by clicking the **Log in** button next to the name of our store.

2. After successfully logging in to our store, in the sidebar, under **Sales channels**, click on **Online Store** to expand the additional sections of the menu and subsequently click on the **Pages** link, redirecting us to the **Pages** section of our store:

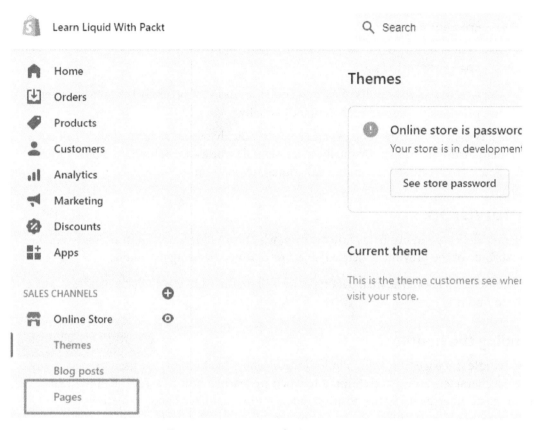

Figure 2.1 – Accessing the Pages section

3. Once inside, since our store is new and has no content, the only thing that we will see is the **Add page** button. Otherwise, we would see a list of all the pages that we have access to in our store. To proceed with creating our first page, click on the **Add page** button.

4. After initializing the process of creating a new page, we will see the page where we can define the content of the new page, including the name, description, visibility, and which template the page should use. We will only need to enter the title of our page for our current purpose, ensuring that we select the **Visible** option under the **Visibility** section, where the **Content** field can remain empty for now. For our page, we have entered `Learning about the page handle` as the title. We also made sure to select the **Visible** option, so the only thing left is to click on the **Save** button:

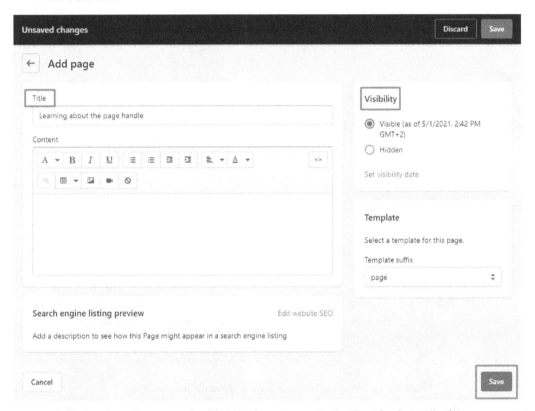

Figure 2.2 – Creating and publishing the new page in the Shopify admin interface

5. Now that we have created our page, we can see the page handle by clicking on the **Edit website SEO** button on the lower part of the window, which will expand and reveal the SEO information for our page, where, among other options, we can see the handle of our page:

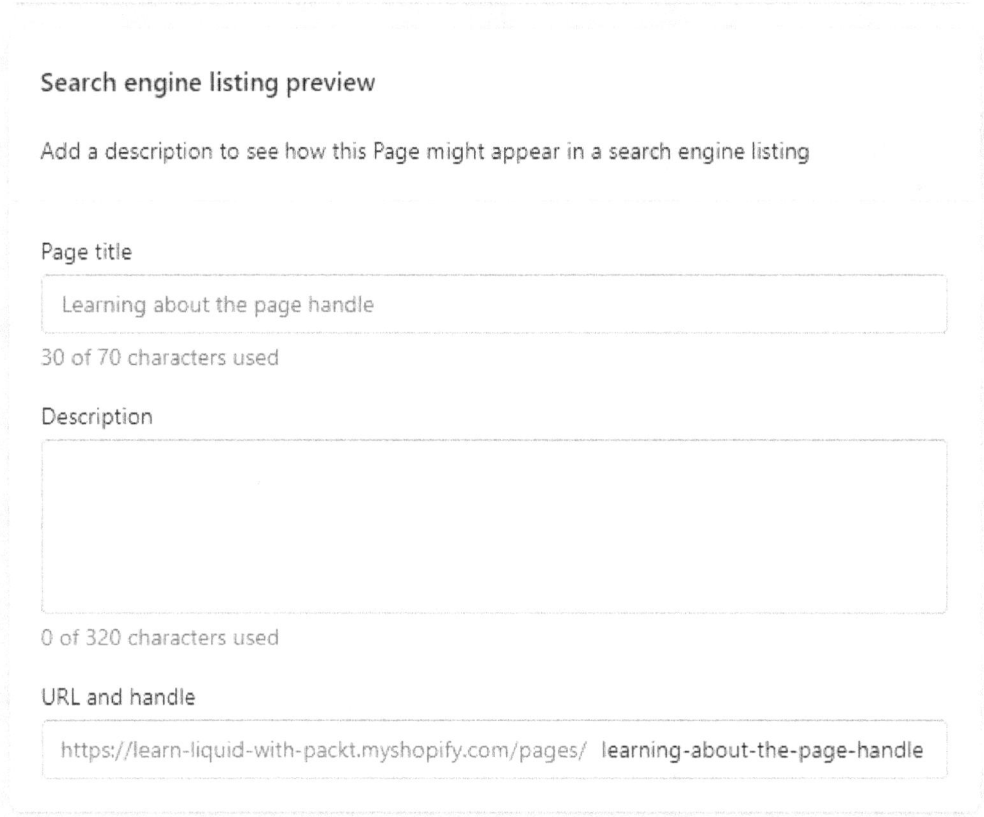

Search engine listing preview

Add a description to see how this Page might appear in a search engine listing

Page title

| Learning about the page handle |

30 of 70 characters used

Description

| |

0 of 320 characters used

URL and handle

| https://learn-liquid-with-packt.myshopify.com/pages/ learning-about-the-page-handle |

Figure 2.3 – Updating the page URL and handle information

As we recall, the handle is a page title in lowercase with hyphens (-) replacing the special characters and empty spaces. In our case, the handle for our page is `learning-about-the-page-handle`.

> **Important note:**
> If for some reason, we were to change the title of the page that we have now created, this would only affect the page title. The handle would remain the same, as changing the page name does not automatically update the handle. The only way to change the handle is to click on **Edit website SEO** and manually edit the handle inside the **URL and handle** field.

Now that we have learned what the page handle is and how to manage it, we can proceed from where we left off with the EmptyDrop data type.

EmptyDrop data type

As we recall, EmptyDrop results from us trying to access an attribute of the previously deleted or disabled object. Note that EmptyDrop, like nil, is not a string with the text EmptyDrop, nor a variable, and as such, it does not have a visual representation.

We can access an object using its handle by pluralizing the name of the object that we are trying to access, followed by either a squared bracket ([]) or dot (.) notation:

```
<h1>{{ pages.learning-about-the-page-handle.title }}</h1> \
<h1>{{ pages["learning-about-the-page-handle"].title}}</h1>
```

While the result will be the same no matter which notation we select, it is important to mention both, as they each have their purpose, which we will cover in the later chapters:

```
<h1>Learning about the page handle</h1>
<h1>Learning about the page handle</h1>
```

We have now learned how to access an object and read its attributes through the page handle. However, *what if we were to go back to our admin interface and disable the previously created page by switching the Visibility option to Hidden?* The result of such an action would be EmptyDrop:

```
<h1></h1>
<h1></h1>
```

EmptyDrop attributes have only one attribute called empty?, which is always truthy. To avoid this issue, we can create a conditional statement to check whether EmptyDrop is empty:

```
{% if pages["learning-about-the-page-handle"].title !=
  empty %}
  <div class="tester">{{ pages["learning-about-the-page-
    handle"].title }}</div>
{% endif %}
```

If the object is not equal to empty, which is always truthy for EmptyDrop, the code inside our conditional will render. Otherwise, if the code is equal to empty, the object we are looking for is empty, and the code will not render.

Controlling whitespace

In the previous sections, we learned how to use conditionals with variable data types to ensure that we always receive a correct value. However, even with our conditionals, the same values can be accompanied by some unwanted whitespaces:

```
Collection info:


{% if collection %}
  The collection's name is {{ collection.title }} !
{% endif %}
```

In the previous example, we have created a conditional that will return `true` if our collection object exists, ensuring that our message will not be incomplete. While our result looks correct, if we were to inspect the page, we would see that things aren't perfect:

```
Collection info:
  The collection's name is Winter Shoes !
```

As we can see from our previous example, we have successfully recovered the collection information. However, we can see a significant number of empty spaces around our message, which results from processing Liquid code. Even though not every Liquid code will output HTML, each line will generate a line for each line of Liquid code by default.

To resolve this problem, we can introduce a hyphen inside our syntax tag as {{- - }} or {%- -%}, which allows us to strip any unwanted whitespace:

```
Collection info:


{%- if collection -%}
  The collection's name is {{ collection.title -}} !
{%- endif -%}
```

With the introduction of the hyphens seen in the previous code block, we have successfully removed the two empty lines that our conditional was rendering:

```
Collection info:
The collection's name is Winter Shoes!
```

Notice how the hyphens we added on both ends within our conditional has removed the whitespace on both sides. However, we have only added one hyphen on the closing bracket of the `collection.title` to remove the empty whitespace on its right side. If we were to add a hyphen to the left side as well, we would have removed the space separating the verb `"is"` and the collection name.

Summary

In this chapter, we covered the basics of Liquid by learning what Liquid is, and how to discern Liquid code by learning how to read and write its syntax. We have gained an understanding of all Liquid logic and comparison operators, which, combined with all Liquid data types we have covered, will help us ensure that we always receive only the data we are expecting. As we progress further, this will become more and more essential.

Lastly, we have learned how to create and access each page individually, which will be of great use to us in the following chapter, where we will be learning more about the objects that we have mentioned on a few occasions throughout the current chapter.

Quiz

1. What type of delimiter should we use if we are expecting an output as a result?

2. What will the result of the following conditional be, and why?

    ```
    {% if collection.all_products_count > "20" %}
        The number of products in a collection is greater
        than 20!
    {% endif %}
    ```

3. What are the two methods to access an item inside an attribute?

4. What is the correct way of accessing an object using its handle?

5. What are the two problems inside the following block of code?

    ```
    {% if customer != nil %}
        Welcome {{- customer.name -}} !
    {% endif %}
    ```

Section 2: Exploring Liquid Core

This section will dive into the Liquid core, which consists of three main features – objects, tags, and filters – which will help us set a solid groundwork for our future learning. We will learn how to use the different types of objects at our disposal, combined with Liquid programming logic, to turn static templates into a fully dynamic store. Lastly, through the power of filters, we will learn how to manipulate various types of data and turn even a simple set of data into complex features.

This section comprises the following chapters:

- *Chapter 3, Diving into Liquid Core with Tags*
- *Chapter 4, Diving into Liquid Core with Objects*
- *Chapter 5, Diving into Liquid Core with Filters*

3

Diving into Liquid Core with Tags

In the previous chapters, we saw some Liquid tags, such as control flow tags, in action. In this chapter, we will learn more about all the different tags we can use to modify our page content dynamically. We will learn about creating variable tags and theme tags and the best way to use them.

Learning about various types of iterations tags and parameters, we will gain the ability to execute blocks of code repeatedly, which will help us write better-quality code. Finally, we will mention some deprecated tags; they still appear in some older themes, so it is essential to know what they do and how to use them.

This chapter covers the following topics:

- Controlling the flow of Liquid (control tags)
- Variable tags
- Iterations tags
- Theme tags
- Deprecated tags

This third chapter will expand our knowledge of logic and comparison operators and different data types by exploring Liquid programming logic.

Technical requirements

You will need an internet connection to follow the steps outlined in this chapter, considering that Shopify is a hosted service.

The dataset used in this chapter, in `*.csv` format, is available on GitHub: `https://github.com/PacktPublishing/Shopify-Theme-Customization-with-Liquid/blob/main/Product-data.csv`.

The code for this chapter is available on GitHub: `https://github.com/PacktPublishing/Shopify-Theme-Customization-with-Liquid/tree/main/Chapter03`.

The Code in Action video for the chapter can be found here: `https://bit.ly/3nP8uwG`

Getting things ready

Before we can proceed, we will need to create some product and collection pages first, which we will be using throughout the following exercises.

Creating the product page

Generally, creating a page or a product page is a straightforward and intuitive process:

1. We can start by navigating to the **Products** section in our sidebar and clicking the **Add products** button, automatically redirecting us to the page to define our product name, description, image, price, and other parameters. We will not go into too many details on managing a product. However, if you would like to read more about this topic, refer to `https://help.shopify.com/en/manual/products/add-update-products`.

2. To avoid creating a significant number of products manually that we will require later, we will use already created products that we can find on GitHub. We have created a `.csv` file that allows us to easily create many products for our development store for this specific purpose. The only thing we need to do is download the file from `https://github.com/PacktPublishing/Shopify-Theme-Customization-with-Liquid/blob/main/Product-data.csv` and import it to our store.

3. Once we have downloaded the file, click on the **Products** link within our sidebar, which will automatically position us in **All Products** after expanding. Click on the **Import** button to trigger a popup and start the process:

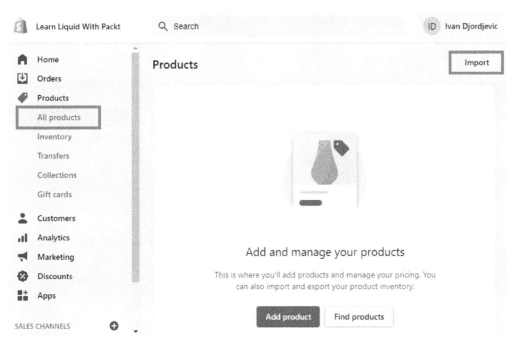

Figure 3.1 – Example of starting the process of importing products

4. After selecting the Product-data.csv file, we can start the process by clicking the **Upload and continue** button:

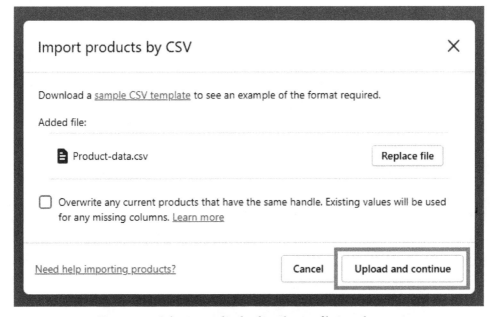

Figure 3.2 – Selecting and uploading the .csv file into the store

5. A few seconds later, we will see another popup previewing one of the products we are about to import. After confirming that the fields contain the correct information, we can finalize the import process by clicking on the **Import products** button:

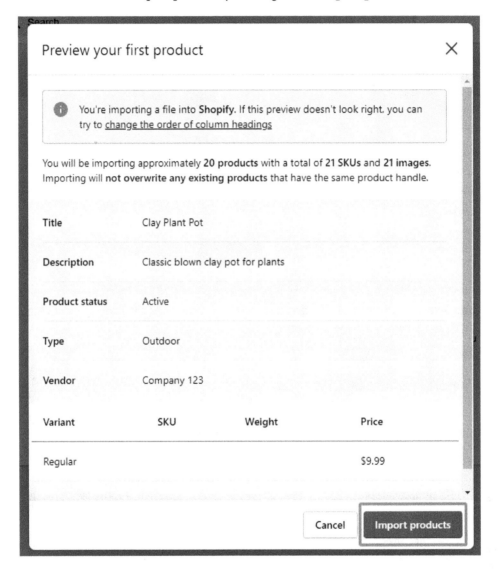

Figure 3.3 – Finalizing the process of uploading the products from the .csv file to the store

Creating the collection page

Now that we have filled our store with products, it is time to create some collection pages that we can then populate with our new products. The process of creating a collection page is as straightforward as that of the product page:

1. We can start the process by clicking on the **Products** link in our sidebar and subsequently clicking on the **Collection** sub-link located in the extended **Products** menu.

2. Once inside, click on the **Create collection** button, automatically redirecting us to the page to define our collection name, description, assign products, and other parameters.

3. Considering that we have two types of products, we will create two collections and assign the same type of products to each collection. After clicking on the **Create collection** button, enter the name Outdoor as the collection name, and set **Collection type** to **Automated** so that we will not have to assign each product manually. The last step is to set **CONDITIONS** such that we should only populate the products whose type equals **Outdoor**:

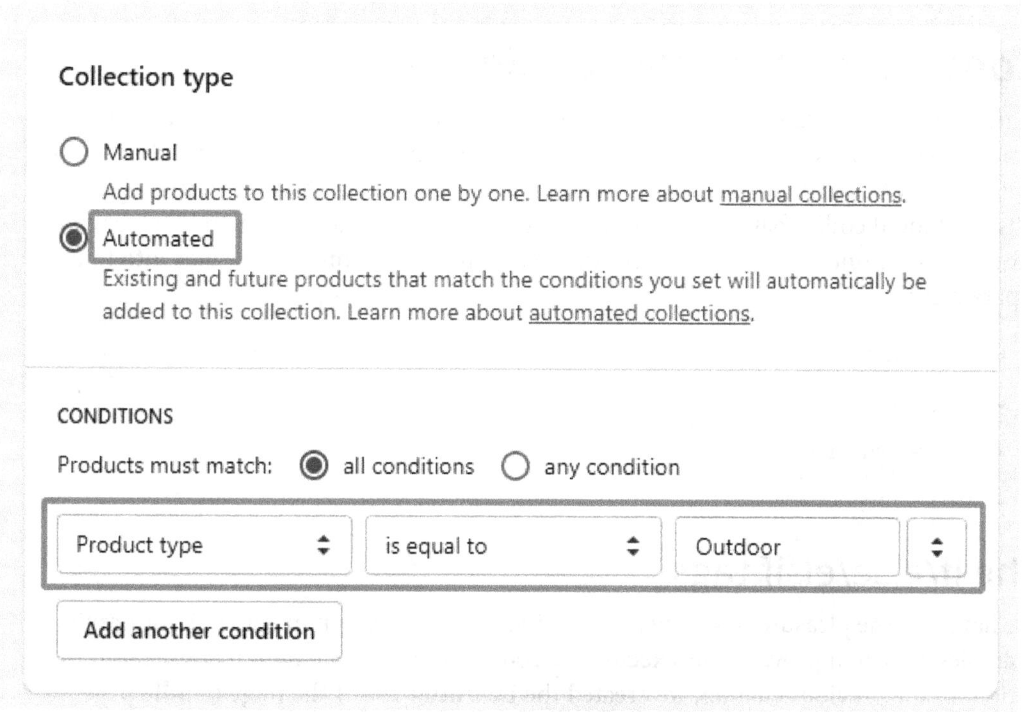

Figure 3.4 – Example of creating a collection and automatically populating it based on the product type

After saving the changes, our collection page will instantly be created and populated with the products that match our condition. Note that before we proceed, we should repeat the preceding steps for the **Indoor** collection, which we will need in the following chapters. If you would like to read more about managing a collection page, refer to `https://help.shopify.com/en/manual/products/collections`.

Updating the navigation menu

For faster navigation, we can include the links to our two collections inside our main menu navigation:

1. We can do this by clicking on **Online store** in our sidebar to expand it and consequently clicking on the **Navigation** link.

2. Once inside, we can add any number of links by opening **Main menu** and clicking on **Add menu item**. If you would like to read more about managing a navigation menu, refer to `https://help.shopify.com/en/manual/online-store/menus-and-links`.

Now that we have finished setting up both our product and collection pages, we can learn about Liquid programming logic.

Controlling the flow of Liquid

In the previous chapters, we saw some control flow tags, such as `if`, `and`, and `or`, in action; now we will dive further into this topic and learn about all the control flow types of tags and how to use them. Control flow tags are a type of Liquid programming logic that tells our Liquid code what to do by allowing us to be selective about which block of code should execute under specific conditions. We can divide the control flow tags into four separate groups:

*   `if/else/elsif`
*   `and/or`
*   `case/when`
*   `unless`

The if/else/elsif tags

We have had the pleasure of seeing the conditional `if` statement in some of our previous examples, which, if proved true, execute the code inside our statement. Let's see it in action. In the previous chapter, we created the **Learning about the page handle** page.

However, let's try and create a new page for this exercise to solidify our knowledge and keep everything concise:

1. Let's start by creating a new page named **Controlling the flow of Liquid**, which we will continue to use going forward. Suppose we need to remind ourselves how to create a new page; we can revisit the previous chapter and consult the *EmptyDrop* subsection, located under the *Understand the types of data* section, where we previously outlined the process of creating a new page.

2. After creating the new page, it is time to edit our newly created page template file. We need to navigate to the **Themes** section, located under the **Online store** area, click the **Actions** button on the duplicate theme we have created, and select the **Edit code** option, which will open our code editor.

3. Once inside the code editor, find the template currently assigned to our page. In our case, the template name is page.liquid, so click on it.

4. Currently, our page consists of two Liquid elements: {{ page.title }}, which generates the title of the page, and {{ page.content }}, which generates the content of our page, along with a few HTML elements. However, *what if our* {{ page.content }} *is empty?* We would end up with an empty div element:

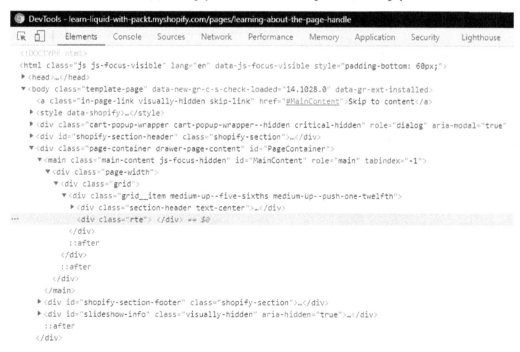

Figure 3.5 – Example of the empty element on a page

To solve this, we can wrap {{ page.content }} along with the div that encapsulates it within a conditional that will check whether the content of the page is empty:

```
{% if page.content != blank %}
  <div class="rte">
    {{ page.content }}
  </div>
{% endif %}
```

We are now sure that we will not see any extra empty strings and elements on our page with the conditional in place.

Let's now try to include some kind of placeholder text instead and let anyone who visits our page know that we will add the page content shortly. This is where the else and elsif statements come to help us. With the addition of else and elsif, we can create multiple conditions to ensure that we execute our code correctly:

1. Let's start by adding the else condition before {% endif %} to execute a different result when our {{ page.content }} is empty:

```
{% if page.content != blank %}
  <div class="rte">
    {{ page.content }}
  </div>
{% else %}
  <div class="rte">
    The page content is currently pending. Please
      check again shortly.
  </div>
{% endif %}
```

With the `else` condition in place, if someone accidentally visits our page, they won't think that we have a broken page, as we will have a message waiting for them:

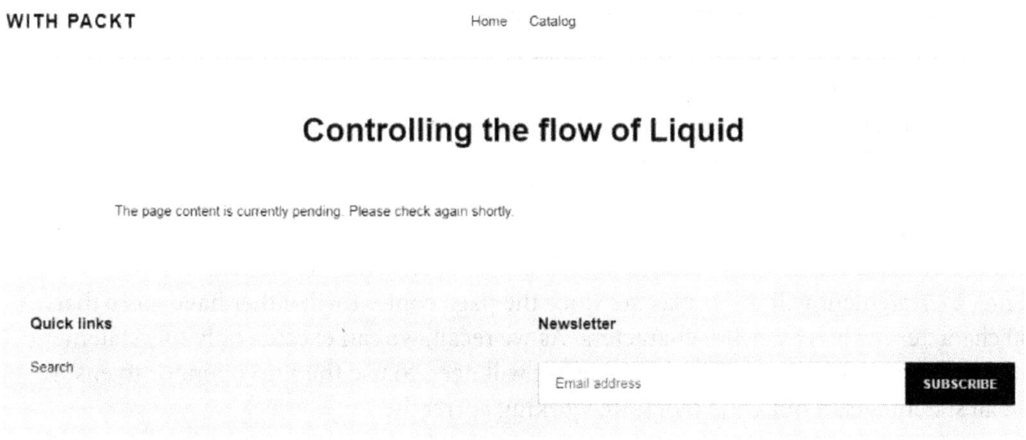

WITH PACKT   Home   Catalog

Controlling the flow of Liquid

The page content is currently pending. Please check again shortly.

Quick links   Newsletter

Search   Email address   SUBSCRIBE

Figure 3.6 – Example of the else condition getting executed

2. Let's try to improve our page by adding a statement to check whether we have started working on our page content and update our notification message accordingly. We can do this by adding an `elsif` statement to check whether our page content has more than 100 characters. However, for this to work, we will also need to modify the existing `if` statement by altering the existing conditional to display the page content if it has `100` or more characters:

```
{% if page.content.size > 100 %}
  <div class="rte">
    {{ page.content }}
  </div>
{% elsif page.content.size < 100 %}
  <div class="rte">
    Our content writers are making the final touches,
      and the page content will be available shortly.
  </div>
{% else %}
  <div class="rte">
    The page content is currently pending. Please
      check again soon.
  </div>
{% endif %}
```

The first conditional statement in our previous example will check whether our page content has more than 100 characters. If it does, it will return true, and our page content will show. The second if statement, which checks whether the page content has less than 100 characters, will only happen if the first statement returns false.

> **Important note:**
> It is impossible to execute multiple statements on different lines within the same block, even if their result was true.

With the two conditionals in place, our page should have enough information for anyone visiting our page. However, if we look closely at our statements, we will see that the {% else %} statement will never execute since the page content will either have more than 100 characters or less than 100 characters. As we recall, we can execute only one statement within a single block. To resolve this issue, we will need to use the and operator to ensure that all statements in our code block are working correctly.

The and/or tags

As we recall from the previous chapter, the and and or operators let us include more than one condition in a control flow tag, allowing us to create complex conditionals.

Using the and operator, we can chain another condition to the existing statement, which will only return true if both the left and right sides of the statement are true. Let's try to include another conditional within our elsif statement that checks whether the page content has any content:

```
{% if page.content.size > 100 %}
  <div class="rte">
    {{ page.content }}
  </div>
{% elsif page.content.size < 100 and page.content != blank %}
  <div class="rte">
    Our content writers are making final touches, and the
      page content will be available shortly.
  </div>
{% else %}
  <div class="rte">
```

```
    The page content is currently pending. Please check
        again soon.
  </div>
{% endif %}
```

With the addition of the second condition, we have ensured that our control flow tag will execute correctly. The first statement will return `true` if the page content has more than `100` characters, the second statement will return `true` if the page content has less than `100` characters and the page is not empty. And finally, if both previous statements return `false`, the code inside the `else` statement will execute.

Similar to the `and` parameter, the `or` parameter allows us to chain another condition to the tag. However, the critical difference is that for a statement with the `and` parameter to return `true`, both the left and right sides of the conditional had to return `true`. We only need at least one condition to return `true` with the `or` parameter, which will make that statement truthy, and the code inside will execute.

The case/when tags

As with the `if`/`elsif`/`else` conditions, `case`/`when` is a type of control flow tag we can use to create a `switch` statement, which allows us to execute a particular block of code only when the returned value is an exact match. We can use `case` to initialize the `switch` statement, and we can use `when` to set the conditions in a specific order that we want them executed in.

Let's return to the `page.liquid` template that we have previously worked on and create a `switch` statement that will check the exact number of characters our page content has and render the appropriate message depending on which statement is true. We can include this functionality above the first `if` statement we have added:

```
{% case page.content.size %}
  {% when 150 %}
    The page has only 150 characters, so this should not
        take much time to read.
  {% when 350 %}
    We have bumped the page to 350 characters, but it
        should not take that much time.
  {% when 1000 %}
    We now have 1000 characters written, and it is going to
        take a few minutes to read everything carefully.
{% endcase %}
```

In the preceding example, we have initialized the `case` tag with `page.content.size`, allowing us to use `when` statements to check whether the number of characters in a page's content is strictly equal to our values. Notice that `when` statements hold no comparing variables. This is because `when` statement accepts only a value as a parameter and will return `true` only if the value is an exact match:

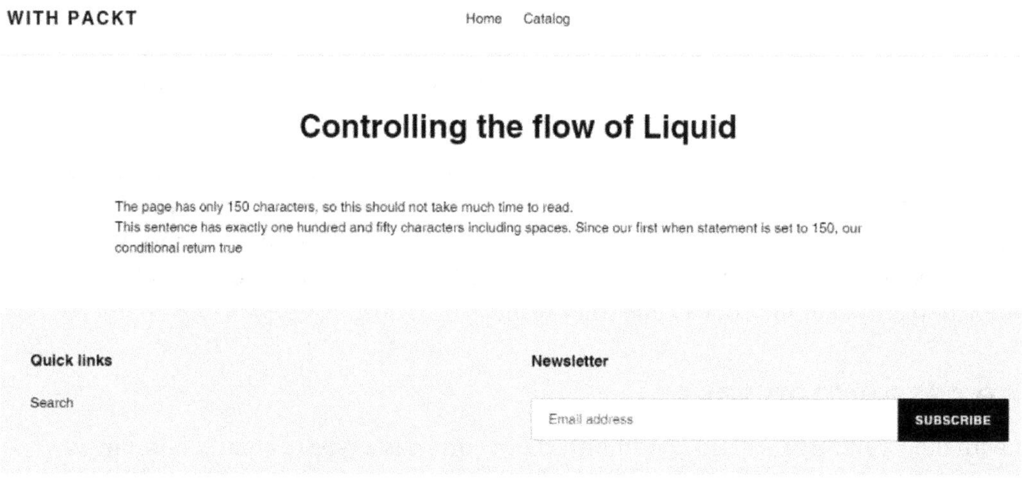

WITH PACKT   Home   Catalog

Controlling the flow of Liquid

The page has only 150 characters, so this should not take much time to read.
This sentence has exactly one hundred and fifty characters including spaces. Since our first when statement is set to 150, our conditional return true

Quick links   **Newsletter**

Search

Email address   SUBSCRIBE

Figure 3.7 – Example of executing the case/when tag

With our control flow tag in place, if we enter `150` characters as our page content, the first `when` statement returns `true`, and as a result, we can see our message. In comparison, it might not look like a powerful tag considering that we can only use it to match the exact values as opposed to the `if/elsif/else` tag. However, the `case/when` tags are a mighty piece of programming logic; we will be using them in the following chapters of the book to create complex functionalities.

As we recall, any code added directly to the template file will get executed on all of the pages using that particular template. And since we have added all of our previous code to the `page.liquid` template, we should ensure that we only execute the previously added code on the **Controlling the flow of Liquid** page. For this, we can use the `unless` tag.

The unless tag

Similar to an `if` tag, which allows us to check the state of a specific condition, the `unless` tag allows us to check if we have not met the condition:

```
{% unless page.title != "Controlling the flow of Liquid" %}
    Unless the page title is Controlling the Flow of Liquid,
        the code inside this conditional will not execute.
{% endunless %}
```

We can add the opening statement of the `unless` tag just above our opening `case` tag and add the closing statement below the closing statement for our `if`/`elsif`/`else` tag:

page.liquid ✕

---

page.liquid Older versions

```
 1 ▾ <div class="page-width">
 2 ▾   <div class="grid">
 3 ▾     <div class="grid__item medium-up--five-sixths medium-up--push-one-twelfth">
 4 ▾       <div class="section-header text-center">
 5            <h1>{{ page.title }}</h1>
 6          </div>
 7
 8          {% unless page.title != "Controlling the flow of Liquid" %}
 9            {% case page.content.size %}
10              {% when 150 %}
11                The page has only 150 characters, so this should not take much time to read.
12              {% when 350 %}
13                We have bumped the page to 350 characters, but it should not take that much time.
14              {% when 1000 %}
15                We now have 1000 characters written, and it is going to take a few minutes to read everything careful
16            {% endcase %}
17
18            {% if page.content.size > 100 %}
19 ▾          <div class="rte">
20              {{ page.content }}
21            </div>
22            {% elsif page.content.size < 100 and page.content != blank %}
23 ▾          <div class="rte">
24              Our content writters are making final touches, and the page content will be available shortly.
25            </div>
26            {% else %}
27 ▾          <div class="rte">
28              The page content is currently pending. Please check again soon.
29            </div>
30            {% endif %}
31          {% endunless %}
32
33        </div>
34      </div>
35 </div>
```

Figure 3.8 – Example of the entire code related to the control flow of Liquid

With the `unless` statement in place, we have ensured that all of our code will only execute on this particular page and will not affect any other page that uses the same `page.liquid` template file.

By learning about all the different types of control flow tags, we have taken a step forward in mastering Liquid programming logic, which will serve as a stepping stone to much more significant and complex things that await us.

In the previous chapter, we have mentioned various data types, such as strings, numbers, and Booleans, and how we can use them. However, what if we wanted to save any of these data types and re-use them on multiple locations without manually updating each line? This is where *variables* come into play.

Variable tags

We can consider variables as data containers to save the various types of information that we want to use later in our code or overwrite as needed. Besides saving the information for later use, variables also allow us to use descriptive text as a label, allowing us to understand what type of information a particular variable contains. We can divide the variable tags into the following four groups:

- `assign`
- `capture`
- `increment`
- `decrement`

The assign tag

The `assign` tag allows us to declare a variable to which we can assign string, number, or Boolean data. We can declare a variable by writing the `assign` keyword followed by the name of the variable we are declaring, followed by the equal sign and the date we assign to that particular variant, and encapsulating it within the curly brace delimiters with percentage symbols:

```
{% assign stringVar = "This is a string!" %}
{% assign numberVar = 2021 %}
{% assign booleanVar = true %}
```

Once we declare a variable, we can call it as many times as we need by encapsulating the variable's name within double curly brace delimiters:

```
{{ stringVar }}
{{ numberVar }}
{{ booleanVar }}
```

Calling our three variables will generate the same type of data that we have initially assigned to each variable:

```
"This is a string!"
2021
true
```

Note that calling a variable using the double curly brace delimiters will return the variable value when called on its own. If we wanted to use a variable within a `for` tag or an `if` statement, we would call the variable only using its name, without the double curly brace delimiters:

```
{% if stringVar == "This is a String!" %}
   The variable content is a string value!
{% elsif stringVar == 2021 %}
   The variable content is a number value!
{% else %}
   The variable content is a boolean value!
{% endif %}
```

We have used our previously declared variable inside a multiple `if` statement to determine our variable's value. Since our variable value is equal to the compared value of the first statement, the first statement will return `true`.

So far, we have learned how to assign a single type of data to a variable, *but what if we wanted to create a variable that will store a combination of a string and a variable?* To achieve this, we will need the help of another type of variable tag called `capture`.

The capture tag

As opposed to the `assign` tag, which only allowed us to capture a single value, `capture` allows us to capture multiple values using its start and end closing tags. We can declare a `capture` variable with a set of curly brace delimiters with percentage symbols around the word `capture` followed by the name of the variable we are declaring:

```
{% assign percent = 15 %}
{% assign deal = "three or more" %}


{% capture promoMessage %}
   If you order {{ deal }} books, you can receive up to a {{
      percent }}% discount!
```

```
{% endcapture %}
```

```
{{ promoMessage }}
```

In the previous example, we initially defined two variables using the `assign` tag, one containing the number value and one containing the string. Once we have declared the two variables, we have again declared a new variable using the `capture` tag. We have included a string message containing both the previously defined variables, and finally, we have called the `promoMessage` variable by wrapping it inside double curly brace delimiters to see the result of our work:

```
"If you order three or more books, you can receive up to a 15%
discount!"
```

As we can see, using the `capture` tag, we successfully created a complex string message, which will come in handy as we progress further with our knowledge.

> **Important note:**
> While the `capture` tag accepts all data types, as we had the chance to see in the previous example, the result of calling the capture variable will always return string data.

Using the `assign` tag, we have learned how to create new variables where we can store a single type of data. Using the `capture` variables, we have learned how to create complex strings using different variables, *but how would we create a variable whose content is a number value that acts as a counter?*

The increment tag

As opposed to the `assign` and `capture` tags, the `increment` tag does not need to be declared first. Instead, we will automatically create the variable as soon as we call it the first time. The `increment` tag allows us to auto-create a variable and increment it every time we call the variable using the `increment` tag. We can call an `increment` variable with a set of curly brace delimiters with percentage symbols around the word `increment` followed by the name of the variable we are creating:

```
{% increment counter %}
```

The starting value of this and any other variable created using the `increment` tag will always be zero.

> **Important note:**
> Calling the `increment` variable will not only automatically generate and
> increment the value starting with zero, but it will also automatically output the
> content of the variable in the template we are working on.

Considering that the `increment` tag automatically outputs the value as soon as the tag is
called, it has a pretty limited usage. The most common use is to auto-generate the unique
numbered identifiers for HTML elements:

```
<div class="product-count">
  <div id="product-item-{% increment counter %}"></div>
  <div id="product-item-{% increment counter %}"></div>
  <div id="product-item-{% increment counter %}"></div>
</div>
```

The result from our previous example will allow us to create a unique ID for each `div`
element, starting with zero for the first occurrence of the `increment` tag and increasing
its value by one for each next occurrence:

```
<div class="product-count">
  <div id="product-item-0"></div>
  <div id="product-item-1"></div>
  <div id="product-item-2"></div>
</div>
```

One more critical aspect of the `increment` tag is that it works independently from
variables created using `assign` or `capture` tags. Let's try to create a variable using the
`assign` tag and see what happens when we try to increment it:

```
{% assign numberVar = 7 %}

{% increment numberVar %}
{% increment numberVar %}

{{ numberVar }}
```

Initially, we created the variable and have assigned it a value of 7, after which we used the `increment` tag to call the variable with the same name twice. Finally, we called the variable that we initially created using the `assign` tag:

```
0
1

7
```

As we can see from our results, even though we have already declared the `numberVar` variable using the `assign` tag and have assigned it a value of 7, the `increment` tag has started incrementing the values starting with 0. They might share the same name, but they are entirely different variables. The `increment` variable will not affect the initially created variable in any way and vice versa.

> **Important note:**
> Note that a variable created using the `increment` tag cannot be called independently without the `increment` tag or used as a logic parameter, as opposed to the case with the `assign` and `capture` tags.

We have now learned how to create an independent variable tag to create a unique element identifier for any number of elements. However, suppose, for some reason, that we needed a variable that outputs the negative values for a large number of elements. For this functionality, we will need to use a different type of variable tag, named `decrement`.

The decrement tag

Besides using a `decrement` keyword to initialize the variable, the `decrement` tag differs from the `increment` tag in two key aspects:

- The first one is that `decrement` allows us to decrease the variable value by one for each occurrence.

- The second one is that the initial value starts with a negative value of minus one.

In the following example, we can see an example of calling the `decrement` tag three times using the same variable:

```
{% decrement numberVar %}
{% decrement numberVar %}
{% decrement numberVar %}
```

In this initial example, we have called the decrement tag three times. Since the decrement variable starts with a negative value, after calling it three times, we will receive the following values:

```
-1
-2
-3
```

> **Important note:**
> Besides the two differences that we have made, decrement shares all the rules and limitations of the increment tag, meaning that the decrement tag works independently of the variables created using the assign and capture methods. We cannot call it independently without initializing the decrement tag.

Variables are a potent tool that, combined with the iteration tags, will bring us one step closer to writing more concise and reusable code.

Iterations tags

Iterations tags are different Liquid programming logic types that allow us to run blocks of code repeatedly. Using iteration tags will save us the time that it would otherwise take us to execute code for each occurrence manually; plus, it will make our code much more concise and readable. To keep the topic concise, we will only mention some of the most used iteration tags and their parameters, which is more important than listing them all as they are all created using similar concepts. We can divide the iteration tags into four separate groups:

- for/else
- jump statements
- for parameters
- cycle

The for/else tags

In the previous chapter, we had the chance to use the for loop in one of our examples when we explained the arrays inside the *Understand the types of data* topic. However, we have not had the chance to explain all the possibilities that the for loop gives us. The for loop is a type of Liquid programming logic that allows us to loop over any code block or array and output the result of that loop for further use.

Previously, we have worked on the `page.liquid` template, but now, we will move on to the `collection.liquid` template.

Let's start by writing a `for` loop tag that will list all of our product names and place them under the `{% section 'collection-template' %}` line:

```
{% for product in collection.products %}
    {{ product.title }}
{% endfor %}
```

In the previous example, we used a `for` tag followed by the `product` variable to loop over the `collection.products` object and return the names of the products currently assigned to our collection, which we can see on our **Indoor** collection page:

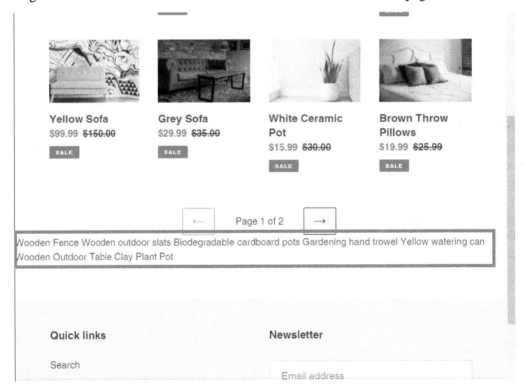

Figure 3.9 – Example of using a for loop to list the names of the products

While our `for` loop did make it work, and we can see our product names listed, this is not of much use since we already have those same products listed. Let's use what we learned in the previous chapter when we talked about different ways of accessing the page handles' objects in the *EmptyDrop* subsection, which we can find under the *Understand the types of data* section.

With our current code in place, we are reading the products of the collection we are currently visiting. Instead, let's try and access the products located in our second collection named outdoor:

```
{% for product in collections["outdoor"].products %}
   {{ product.title }}
{% endfor %}
```

Even though we are currently previewing the Indoor collection, we have now gained a list of products that belong to the Outdoor collection.

jump statements

As their name suggests, jump statements allow us to set the conditions that exclude certain items from our loop or stop our loop when we meet a specific condition. We can split jump statements into the two following groups:

- continue: The {% continue %} statement allows us to skip the current iteration based on the condition that we have set. We can use it to exclude certain products from our for loop by pairing it with an if statement only to accept a product whose price is lower than $100.00. Otherwise, we should exclude the product from the loop:

```
{% for product in collections["outdoor"].products %}
   {% if product.price > 10000 %}
      {% continue %}
   {% else %}
      {{ product.title }}
   {% endif %}
{% endfor %}
```

With the if statement in place, any product iteration whose price is higher than $100.00 will trigger our {% continue %} jump statement, consequently excluding that product from our iteration. Otherwise, any product that fails that statement will render the product name on our page.

- break: The {% break %} statement, on the other hand, allows us to stop the loop when we met a specific condition:

```
{% for product in collections["outdoor"].products %}
    {% if product.price > 10000 %}
        {% break %}
    {% else %}
        {{ product.title }}
    {% endif %}
{% endfor %}
```

When we used the {% continue %} statement, the result we have received from our loop was that we had excluded any product whose price was lower than $100.00 from the loop result. However, in our latest example, instead of excluding the products that match our condition, {% break %} will cause the loop to stop and not perform any other iteration the moment it finds the first occurrence that matches our condition.

For example, with {% continue %}, our loop returned six products that match our condition. However, when we replace {% continue %} with {% break %}, our loop returned zero results. Since the first product inside the Outdoor collection, whose handle we are using, has a price of $200.00, we have met our condition on the first iteration, which caused {% break %} to stop the iteration, preventing all other iterations.

The for parameters

In addition to jump statements, we also have various parameters that we can use further to define the loop's limits and workflow. We can split the parameters that we can use in conjunction with the for loop into the four following groups:

- limit
- offset
- range
- reversed

limit

As its name suggests, the limit parameter limits how many iterations our loop should perform. We can add limit parameters at the end of the opening for loop tag, followed by a colon, followed by a number value.

We usually use limit parameters when we require a particular number of iterations without implementing any condition. This is because the limit parameter only measures the number of iterations, not whether the number of iterations matches the statement we have in place.

Let's start by adjusting our previous example by replacing {% break %} with {% continue %} and adding the limit parameter with the value of 4, which is the maximum number of products that we are looking to get as a result:

```
{% for product in collections["outdoor"].products limit: 4 %}
  {% if product.price > 10000 %}
    {% continue %}
  {% else %}
    {{ product.title }}
  {% endif %}
{% endfor %}
```

In our previous example, before adding the limit parameter, our for loop returned six product iterations whose price was lower than $100.00. With the addition of the limit parameter, the number of iterations returned was 3, even though we set our limit to 4. *Why?*

Similar to the {% break %} statement we mentioned earlier, the limit parameter's aim is to limit the number of iterations to the assigned value. Let's open our Outdoor collection, whose handle we are using to loop over. We will notice that the first product in that collection has a price of $200.00, which triggered the if statement, subsequently triggering the {% continue %} statement, which excluded the first product.

As a result, our for loop did not print out that product name. However, the limit parameter still counts this as one iteration, meaning that it will perform three more iterations. Since the following three products' prices were lower than $100.00, the three products' names were returned by our for loop before stopping. For this reason, the limit parameter is usually only used in a loop without any statements. Otherwise, we risk not meeting the number of iterations we intended, or any at all, since the result can also be zero iterations if none matches the statement that we have set.

The offset parameter

The offset parameter allows us to delay the start of the for loop by starting the loop from the specific index. We can add an offset parameter at the end of the opening for loop tag, followed by a colon, followed by a number value.

Let's take our previous example and try replacing limit with the offset parameter and setting its value to 4:

```
{% for product in collections["outdoor"].products offset: 4 %}
  {% if product.price > 10000 %}
    {% continue %}
  {% else %}
    {{ product.title }}
  {% endif %}
{% endfor %}
```

Our for loop has automatically skipped the first four product iterations and has started iterating the products at index number 5, resulting in our for loop returning three product iterations. *However, what if we also include the limit parameter after the offset parameter and set its value to 1?*

```
{% for product in collections["outdoor"].products offset: 4
  limit: 1 %}
  {% if product.price > 10000 %}
    {% continue %}
  {% else %}
    {{ product.title }}
  {% endif %}
{% endfor %}
```

While at first, it might look like the previous example will not work, this is a fully functional loop with valid parameters. As we have previously mentioned, the limit parameter allows us to limit a for loop to a specific number of iterations. With two parameters added, our for loop will start its iteration at index 5. It will cover one iteration as per the value set with the limit parameter, and then it will stop the for loop, regardless of whether the one iteration managed to get past our if statement.

The range parameter

The `range` parameter offers us similar functionality to the `offset` parameter. The critical difference is that with `range`, we can assign both starting and ending index positions. We can add a `range` parameter at the end of the opening `for` loop tag by writing opening brace delimiters followed by the starting value, followed by two dots, and finally followed by end value and closing brace delimiters:

```
{% for item in (3..5) %}
    {{ item }}
{% endfor %}
```

As we recall, the `offset` parameter does not include the starting position index within the loop. Instead, it starts its loop from the next position, where the range is inclusive for both starting and ending position values. The result of our `for` loop example with the `range` parameter set to `(3..5)` would result in the following:

```
3  4  5
```

Besides accepting the number as its value, as we were able to see in our previous example, `range` also allows us to place variables and objects as both its starting and ending values:

```
{% assign start = 3 %}
{% for item in (start..collections["outdoor"].all_products_
    count) %}
    {{ item }}
{% endfor %}
```

The result of our `for` loop example would return all iterations starting at index 3, all the way to the number of products our collection has. With the ability to assign variables and objects as our starting or ending points, we can now create reusable `for` loops resulting in more concise and better-quality code.

The reversed parameter

The final parameter on our list is `reversed`, which, as its name suggests, allows us to reverse the order of iteration. The `reversed` parameter does not have any value representation, and we can include it at the end of the opening `for` loop tag:

```
{% for item in (3..5) reversed %}
   {{ item }}
{% endfor %}
```

In the previous example, we have added the `reversed` parameter as a secondary parameter next to our `range` parameter, resulting in our tag performing iteration in reverse order:

```
5  4  3
```

As we have had the chance to see for ourselves, parameters are a powerful addition to the `for` tag, which we can use to achieve the necessary type of iteration and get only the results we require.

The cycle tag

`cycle` is another powerful tag that we can only use in combination with the `for` tag to loop over a group of strings and output them to each iteration in the specific order in which they were initially defined. We can define the `cycle` tag by opening curly brace delimiters with a percentage symbol, followed by the word `cycle`, followed by any number of strings separated by commas. Finally, we close the `cycle` tag with closing brace delimiters and a percentage symbol:

```
{% for product in collection.products limit: 4 %}
   <div class="product-item {% cycle "first", "", "", "last"
     %}"></div>
{% endfor %}
```

In the previous example, we have included a `cycle` tag with four different strings inside a `for` loop, which we have limited to four iterations. By implementing the `cycle` tag, we have ensured that our first `div` element will receive a class first, the fourth element will receive a `last` class, while the two elements in between will receive no classes as the two strings for those positions are empty:

```
<div class="product-item first"></div>
<div class="product-item"></div>
<div class="product-item"></div>
```

```
<div class="product-item last"></div>
```

From our result, we can see that the `cycle` tag is quite a valuable tool that we can use to pass data to a specific iteration within a loop in an order that we have defined. *However, what would happen if we removed the limit from our loop or increased it to have nine iterations instead?*

```
{% for product in collection.products limit: 9 %}
   <div class="product-item {% cycle "first", "", "", "last"
      %}"></div>
{% endfor %}
```

In our previous example, we had the exact same number of iterations as the number of strings inside the `cycle` tag. In our newest example, the number of iterations is higher than the number of strings, which means that the cycle will reset and start applying strings again, in order, for as many iterations as needed:

```
<div class="product-item first"></div>
<div class="product-item"></div>
<div class="product-item"></div>
<div class="product-item last"></div>
<div class="product-item first"></div>
<div class="product-item"></div>
<div class="product-item"></div>
<div class="product-item last"></div>
<div class="product-item first"></div>
```

So far, we have learned the basic use of the `cycle` tag and that the tag will continue to output strings for as many iterations as possible, which is quite helpful if we have a single `cycle` tag, *but what if we have two or more?*

```
{% for product in collection.products limit: 6 %}
   <div class="product-item {% cycle "first", "", "", "last"
      %}"></div>
{% endfor %}
```

```
{% for product in collection.products limit: 4 %}
   <div class="product-item {% cycle "first", "", "", "last"
      %}"></div>
{% endfor %}
```

In the previous example, we have created two separate `for` loops, the first one limited to six iterations and the second limited to four. As we have just learned, no matter how many strings we have defined inside the `cycle` tag, the tag will continue to output strings for as many iterations as needed:

```
<div class="product-item first"></div>
<div class="product-item"></div>
<div class="product-item"></div>
<div class="product-item last"></div>
<div class="product-item first"></div>
<div class="product-item"></div>


<div class="product-item"></div>
<div class="product-item last"></div>
<div class="product-item first"></div>
<div class="product-item"></div>
```

As we can see from the results, the first `for` loop has created six iterations in the exact order that we intended. However, the second `for` loop did not produce the same results. Instead, it continued to output the strings starting from the exact position where the previous `cycle` tag stopped.

This type of behavior is logical considering that Liquid currently cannot differentiate between different types of `cycle` tags. However, we can easily resolve this by introducing a parameter called the **cycle group**.

The cycle group parameter allows us to separate `cycle` tags by ensuring that each cycle will output strings starting with position one, regardless of whether or not we have already used a `cycle` tag on the same page. After the `cycle` keyword, we can include the cycle group parameter by adding the string name encapsulated by parentheses, followed by a colon:

```
{% for product in collection.products limit: 6 %}
  <div class="product-item {% cycle "group1": "first", "",
    "", "last" %}"></div>
{% endfor %}


{% for product in collection.products limit: 4 %}
  <div class="product-item {% cycle "group2": "first", "",
    "", "last" %}"></div>
{% endfor %}
```

The introduction of the cycle group parameter ensures that each `cycle` tag functions independently:

```
<div class="product-item first"></div>
<div class="product-item"></div>
<div class="product-item"></div>
<div class="product-item last"></div>
<div class="product-item first"></div>
<div class="product-item"></div>

<div class="product-item first"></div>
<div class="product-item"></div>
<div class="product-item"></div>
<div class="product-item last"></div>
```

We can see from our results that all iterations have received the string initially intended for them even though we have more than one `cycle` tag.

Besides the `for` and `cycle` tags, we have one more type of iteration tag called `tablerow`, which works similarly to the `for` tag. The only difference between the two tags is that `tablerow` returns results formatted as an HTML table. To keep the book concise, we will not be covering the `tablerow` tag or its parameters in this book. However, if you would like to read more about it, refer to `https://shopify.dev/docs/themes/liquid/reference/tags/iteration-tags`.

Theme tags

Theme tags are a special type of tag that give us specific control over both *un-rendered and rendered code*. Using the various types of theme tags that we have at our disposal, we can create a different type of HTML markup for specific templates that is essential for creating a form that will allow customers to purchase products from our store. Additionally, they allow us to select different theme layouts to use for different pages, define sections or snippet files that we can use to create reusable blocks of code, and many other things. We can divide theme tags into the following groups:

- `layout`
- `liquid` and `echo`
- `form`
- `paginate`

- render

- raw

- comment

The layout tag

As we recall from the first chapter, when discussing the Layout directory, we mentioned the importance of the theme.liquid file as it is in this file that we will render all of our files and templates. It is in this file that we arrange the general layout of our pages.

Our pages currently consist of three key elements: **header**, **main content**, and **footer**. Let's say we wanted to remove the header and footer sections from our product pages, or at the very least to replace them with a different set of headers and footers. *How would we do this?*

1. The first step to achieve this would be to create an alternate layout file that our theme product pages will use instead. We can do this by expanding the **Layout** directory in our code editor and clicking on the **Add a new layout** button, which will trigger a popup to select the type of layout page we are looking to create and select the name for our new page.

2. Within the drop-down menu, we will select the **theme** option as the layout type, and within the following field, we will enter the word alternate as our filename. Once we have finished selecting our layout file type and name, click on the **Create layout** button to finalize the process:

Figure 3.10 – Example of creating a new layout file

With this, we have successfully created a new layout file named `theme.alternate.liquid`. Suppose we tried to compare this file with our original `theme.liquid`; we would see that they are exactly the same. The reason for this is that when we create a new page, layout, or template, Shopify will not create a new page entirely, but it will copy a default page that our theme is already using. For example, if we were to expand the **Templates** directory and create a new page template, the new page template would contain all of the changes we had previously made to the `page.liquid` file.

Now that we have created a new layout file, we will assign it to our product pages. We can do this by expanding the **Templates** directory and clicking on the `product.liquid` file where we can define the layout for this particular template, which we should include on the very first line of the file. We can define the `layout` tag with a set of curly brace delimiters with percentage symbols around the word `layout`, followed by the name of the layout file, which we should encapsulate within parentheses:

```
{% layout "theme.alternate" %}
```

If we were to navigate to the **theme.alternate** template now and delete `{% section 'header' %}` and `{% section 'footer' %}`, the header and footer sections would no longer be visible on our product page. However, they would still be visible on pages that use the original `theme.liquid` layout.

> **Important note:**
>
> Besides the name of the layout file, the `layout` tag also accepts the `none` keyword, without parentheses, as its value, in which case the page will use no layout file. In this case, the particular page using this particular template will have no access to any code or file, including `.css`, `.js`, or others that we initially loaded through the `theme.liquid` file.

Any template file that does not contain the `layout` tag will, by default, use the original `theme.liquid` layout as a failsafe.

The liquid and echo tags

The `liquid` tag is one of the newest additions to Liquid, and it is quite a powerful one as it allows us to write multiple tags within one set of delimiters, making our code a lot easier to read and maintain. We can define the liquid tag with a curly brace delimiter and a percentage symbol on the left side of the word liquid followed by as many lines of `Liquid` code as we need. Note that we only need to define an opening `liquid` tag, where the `liquid` closing tag is automatically closed.

The echo tag is an addition to the liquid tag that allows us to output an expression that initially we had to wrap inside double curly brace delimiters. We can define the echo tag by simply writing an echo keyword, followed by any expression that we are using.

Let's try and use our liquid and echo tags to refactor the code that we had previously added to our collection template when we learned about the different types of iterations.

Here, we can see the code that we initially created on our collection page:

```
{% for product in collections["outdoor"].products offset: 4 %}
  {% if product.price > 10000 %}
    {% continue %}
  {% else %}
    {{ product.title }}
  {% endif %}
{% endfor %}
```

We can see the same code in the following example after refactoring using the liquid and echo tags:

```
{% liquid
for product in collections["outdoor"].products offset: 4
  if product.price > 10000
    continue
  else
    echo product.title
  endif
endfor %}
```

After refactoring our code, we can see that we have removed all curly brace delimiters with percentages and replaced all double curly brace delimiters with the echo tag. The only curly brace delimiters with percentage symbols left are the ones that belong to the liquid tag, encapsulating the entire code.

> **Important note:**
> While liquid and echo tags are pretty powerful, we should not use them regularly, as while they help us write multiple tags within one set of delimiters, they are not so kind when it comes to working with strings. The echo tag forces us to wrap every string within the parentheses, making it almost impossible to use non-string data values inside the string.

The form tag

The form tag allows us to automatically output various types of HTML `<form>` along with the required `<input>` elements, depending on the type of the form tag we have called. We can define the form tag within a pair of curly brace delimiters, followed by the form keyword, followed by the form's name encapsulated by parentheses, and lastly, followed by the closing endfor tag surrounded by curly brace delimiters with percentage symbols:

```
{% form "product", product %}
{% endform %}
```

In the previous example, we have selected to generate the product form tag by using the product keyword within the parentheses. However, notice that following the comma, we also have another product keyword without parentheses. Certain forms, such as product forms that we are currently using, require us to pass them a parameter to generate the proper content. In this case, that parameter is the product object.

The most common use for this is on a product page. However, we can also use the form tag within a collection page or anywhere else where we have access to the product object. Let's try to include a form tag within the loop that we previously created on our collection page.

Since we will be dealing with a mix of strings and non-strings data, let's first undo the liquid and echo changes that we previously made. We can do this by clicking on the **Older versions** button, located between the file's name and the **Delete file** button:

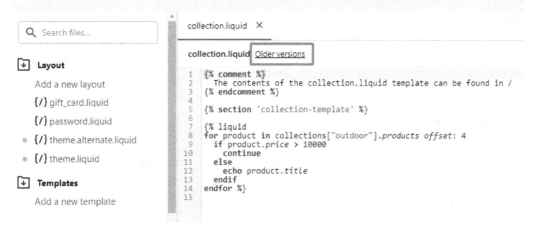

Figure 3.11 – Example of reverting changes in the file

Clicking the **Older versions** button will reveal the drop-down menu where we can see the date and time of all the changes that we have made within this particular file. Each date represents the last time that we pressed the **Save** button to save our changes. Selecting any option will automatically revert our code to that particular point.

After reverting our changes, we are ready to add the `form` tag to our `for` loop:

```
{% for product in collections["outdoor"].products offset: 4 %}
  {% if product.price > 10000 %}
    {% continue %}
  {% else %}
    {% form "product", product %}
      {{ product.title }}
    {% endform %}
  {% endif %}
{% endfor %}
```

By introducing the `form` tag within our `for` loop, we have generated an HTML form tag that we can see by inspecting the string elements that we see on our `Indoor` collection page. While the `form` tag is functional, we still lack one key ingredient to make our form usable, which we will be learning in the next chapter.

We can see that our `form` tag already has quite a few attributes by inspecting our collection page. *However, what if we wanted to add some of our own or modify the existing ones?* We can do this by simply adding the attribute name, followed by a comma, followed by the attribute's value:

```
{% for product in collections["outdoor"].products offset: 4 %}
  {% if product.price > 10000 %}
    {% continue %}
  {% else %}
    {% capture productId %}productId-{% increment counter
    %}{% endcapture %}
    {% form "product", product, id: productId, data-
    location: "collection" %}
      {{ product.title }}
    {% endform %}
  {% endif %}
{% endfor %}
```

In the previous example, we have included a data attribute in our form using a string. However, we have also modified the existing ID value using the previously defined variable.

Notice that while we can directly include the variable within the attribute value, we cannot mix the string and variable within the `form` tag itself. For this reason, we have previously captured the combined string and `increment` tag using the `capture` tag and called it the attribute value using the variable name.

Besides the ability to pass an object as a `form` parameter, the `form` tag provides us with one special parameter named the `return_to` parameter. By default, when we submit the product form, we are usually redirected to the cart page. However, with the `return_to` parameter, we can select the page we should land on after we submit the form:

```
{% form "product", return_to: "back" %}
{% endform %}
```

In our previous example, we have assigned the `back` value as a string for the `return_to` parameter, which will automatically return us to the same page we were on before submitting the form. Besides the `back` string, we can also use a relative path or a `routes` attribute to generate a dynamic URL to the page we should land on once we submit the form.

There are many types of form tags that we can use. However, as they are all created using the same format, we will only cover one example. For more information on different types of `form` tags and their parameters, refer to `https://shopify.dev/docs/themes/liquid/reference/tags/theme-tags#form`.

The paginate tag

Previously, we have learned all about iteration tags, which we can use to repeatedly execute a code block. However, iteration tags have a limitation that means they can only output a maximum of 50 results per page. For any higher number of iterations, we would need to use the `paginate` tag to split the results across multiple pages.

The `paginate` tag must always wrap around a `for` tag that is looping over an array to be able to split the content into multiple pages. We can see an example of a `paginate` tag wrapping a `for` loop here:

```
{% paginate collection.products by 5 %}
  {% for product in collection.products %}
  {% endfor %}
{% endpaginate %}
```

As we can see from our example, the `paginate` tag contains the `paginate` keyword, followed by the object that returns an array of products in the collection, followed by the by parameter and the number value. Depending on the number value that we include with the by parameter, which can go from one to fifty, the `paginate` tag will know the maximum number of items it should present per page. Inside the `paginate` tag, we can access attributes of the `paginate` object. To keep the book concise, we will not be going into detail regarding the `paginate` object. For detailed information, refer to `https://shopify.dev/docs/themes/liquid/reference/objects/paginate`.

The render tag

The `render` tag allows us to render the content of a snippet file to a place of our choosing. As we recall from *Chapter 1*, *Getting Started with Shopify*, snippet files allow us to re-use repetitive pieces of code over **Templates/Sections** by referencing their names:

```
{% render "snippet-name" %}
```

As we can see from our example, we only need to use the `render` keyword, followed by the name of the snippet file, and we will render any content within our snippet file. By using the snippet file, we not only make our code more readable, but we also make it easier to maintain by re-using the same block of code.

> **Important note:**
> After rendering the snippet, we do not automatically have access to the variables created within the parent element. Additionally, the parent element cannot access the variables defined within the snippet.

Let's create a new snippet file named `collection-form`, where we will move the entire content of the `else` statement from the `for` loop we created within our collection page:

```
{% for product in collections["outdoor"].products offset: 4 %}
    {% if product.price > 10000 %}
        {% continue %}
    {% else %}
        {% render "collection-form" %}
    {% endif %}
{% endfor %}
```

While the preceding code looks correct, we would end up with an error if we tried to execute it. *Why?*

As we recall, a snippet file does not have access to the variables defined in the parent element. In this case, that variable is the product we have defined in the `for` loop, which we call inside the snippet as the `form` parameter. To resolve this, we will need to pass the variable to the snippet as a parameter:

```
{% for product in collections["outdoor"].products offset: 4 %}
  {% if product.price > 10000 %}
    {% continue %}
  {% else %}
    {% render "collection-form", product: product %}
  {% endif %}
{% endfor %}
```

We have created a parameter named `product`, to which we have assigned the initially created variable. We could have given our parameter any name we liked, for example, `collection_product`. However, in that case, we would also have to update the `product` parameter inside the `form` tag from `product` to `collection_product` and `product.title` to `collection_product.title`, currently located within our snippet file, to match the new object keyword.

Once we have assigned the variable to a snippet via the parameter, we can modify the variable value inside the snippet independently from the value in the parent template. Even if we overwrite the variable value within the snippet, the variable will keep its value in the parent template. Note that we can pass as many variables as we may need to our snippet using its parameters. *However, what if we wanted to pass an object to our snippet?*

As with variables, we can pass an object to a snippet file using the `with` and `as` parameters. However, we are limited to only one object that we can pass as a parameter:

```
{% render "snippet-file" with collections["collection-handle"]
as featured_collection %}
```

In our previous example, we have used `render` parameters to pass the `collection` object to the snippet file, where we can access it using the `featured_collection` keyword.

The final parameter that we can use with the `render` tag is the `for` parameter. Using the `for` and `as` parameters, we can render a snippet for each occurrence. Let's try to refactor the code on our collection page by moving it inside the `collection-form` snippet and rendering the products using the `for` parameter. However, note that we will not be able to copy the parameters that we currently have on the original `for` tag, as `render` does not accept additional parameters. Additionally, we will need to remove the `capture` and `productId` parameter from the form tag, as the `increment` tag does not work with render for parameter.

The first thing we need to do is to move the `if` and `else` statements inside the snippet file and place them accordingly:

```
{% if product.price > 10000 %}
    {% continue %}
{% else %}
    {% form "product", product, data-location: "collection" %}
        {{ product.title }}
    {% endform %}
{% endif %}
```

The only thing left to do is to add the additional parameters to the `render` tag. In the following example, we have included the `for` and `as` parameters with the `render` tag, which allowed us to iterate the `collections["outdoor"].products` array and render the snippet for each product iteration that our collection has. Additionally, we pass the `iteration` object as a parameter to our snippet, which we can use inside:

```
{% render "collection-form" for collections["outdoor"]
.products as product %}
```

Since we are already using the `product` keyword inside our snippet file, we successfully pass the object to both the product form parameter and product title. If we run our code now, we will see that the results are exactly the same as they were, but our code looks a lot cleaner.

The raw tag

The `raw` tag allows us to output unparsed Liquid code directly on the page. We can use the `raw` tag by wrapping the content we want unparsed with `raw` and `endraw` tags:

```
{% raw %}
   The name of our collection is {{ collection.title }}.
{% endraw %}
```

In our previous example, we have created a message to update the collection name dynamically. However, since we have wrapped the entire message inside the `raw` tag, Liquid will not process it. Rather, it will return it to us in the exact same way:

```
The name of our collection is {{ collection.title }}.
```

This type of functionality can be pretty helpful, especially when dealing with conflicting syntaxes, such as Handlebars.

The comment tag

As its name suggests, the `comment` tag allows us to leave a comment within our Liquid template files. Any text located between the opening and closing comment tags will not render inside the Liquid template files at all:

```
We are currently learning about the comment tags. {% comment
%} Comment tags are a smart way to comment on our code. {%
endcomment %}
```

Since we have wrapped the second part of our message within the `comment` tag, Liquid will render only the first part:

```
We are currently learning about the comment tags.
```

Comment tags are pretty helpful, as they allow us to leave the necessary information within our files without polluting the DOM with visible comments.

Deprecated tags

Deprecated tags are Liquid tags that are considering as outdated, and we should no longer use them in our developing process. However, we may still encounter them inside some of the older themes, so it is important to recognize them and know what they do.

The only tag that Shopify has deprecated is the `include` tag, which works similarly to the `render` tag, which we have previously covered:

```
{% include "snippet-name" %}
```

The only key difference between the two is that when rendering a snippet using the `include` tag, the code inside our snippet can automatically access and modify the variables within its parent template. This not only creates a lot of performance issues but also makes our code a lot harder to maintain, which is why Shopify replaced it with the `render` tag.

Summary

In this third chapter, we have learned about the Liquid programming logic that allows us to select which code block should execute under specific conditions. We have gained an understanding of variable and iteration tags and how we can use them in combination with different programming logic to execute a block of code repeatedly and recover only specific results.

Lastly, we have learned about the different types of theme tags and how we can use them to output a template-specific HTML markup. By learning how to use snippets to make our code more readable and maintainable, we have already taken a step forward in writing better-quality code. This will be especially important in the following chapter, where we will be diving further into Liquid core and learning more about Liquid objects and their attributes.

Questions

1. What parameters should we use inside a `for` loop if we want to show a maximum of seven iterations while also skipping the first three iterations?

2. What types of data can we assign to a variable created using the `capture` tag?

3. What are the two problems in the following block of code?

```Liquid
Liquid for product in collections["outdoor"].products
  if product.price > 10000
    continue
  else
    product.title
  endif
endfor
```

4. What approach should we take to modify an HTML-generated product form by replacing the existing class attribute with a combination of a string and a variable?

5. What parameter should we use if we want to pass an object from the parent element?

4
Diving into Liquid Core with Objects

In the previous three chapters, we have been referencing objects. However, in this chapter, we will learn about objects, their attributes, and some of the best ways to use them. By learning about objects, we will finalize some projects that we have started and work on new projects to develop our knowledge further.

We will cover the following topics in this chapter:

- Working with global objects
- Improving the workflow with metafields
- Content and special objects

Upon completing this chapter, we will understand what content objects are, why they are mandatory, and how to use them, which is the first step in creating future templates. We will have also learned more about the global objects that we have been referencing up until now.

However, due to the significant number of global objects and attributes they offer us, we will not cover all of them. Instead, we will only explain some of the essential global objects and attributes, which will set us on the right path to understanding objects entirely. By learning about objects, we will also learn to understand metafields, which allow us to store and dynamically output additional data on our store. Lastly, we will learn about the special objects that will help us output some helpful functionalities on our store.

Technical requirements

While we will explain each topic and have it presented with the accompanying graphics, we will need an internet connection to follow the steps outlined in this chapter, considering that Shopify is a hosted service.

The code for this chapter is available on GitHub at `https://github.com/PacktPublishing/Shopify-Theme-Customization-with-Liquid/tree/main/Chapter04`.

The Code in Action video for the chapter can be found here: `https://bit.ly/3u7hqyB`

Working with global objects

We referenced objects and their attributes in the previous chapter. *But what exactly are objects?*

These objects, or so-called **liquid variables**, are elements that allow us to read the content defined in our backend and dynamically output it to help us create better programming logic. We can output the data by combining the objects and attributes encapsulated by *double curly braces*. An example of the global object that we were using in the previous chapter would be `{{ collection.title }}`, where the `collection` keyword would be our object and `title` would be the attribute.

We can reference these global objects inside any file by directly visiting the page whose content we are looking to recover and calling the object, or manually invoking the object for the specific page using its handle and combining it with the variable tags. Let's see this in action.

As you may recall from the previous chapter, while we were working on the project on our indoor collection page, we initially used `collection.title` to recover the collection's name:

```
{{ collection.title }}
```

When used inside the collection page, the preceding code will provide us with the data we are looking for. However, *what if we wanted to access the* indoor *collection object while visiting the outdoor collection page?*

This is where our knowledge of accessing the page object using the page handle, which we covered in *Chapter 2, The Basic Flow of Liquid*, in the *EmptyDrop* subsection of the *Understanding the types of data* section comes to help.

We can access the object using its handle by pluralizing the object's name we are trying to access, followed by either a squared bracket ([]) or dot (.) notation:

```
{% assign customCollection = collections["indoor"].title %}
{% assign customCollection = collections.indoor.title %}
```

By defining a variable with the object of the indoor collection, we can output the data for that collection on any page by simply invoking the object, followed by the attribute that we need:

```
{{ customCollection.title }}
```

While our object name differs from the original object name, it will allow us to output the same information compared to when we used {{ collection.title }} inside the collection template.

> **Important note:**
> In the previous example, we created a variable using the customCollection name. However, note that you can create this variable using any name of your choice, including the collection itself. While the collection keyword is not reserved, we should pay close attention when using the keywords already in use as this can cause unexpected results.

Considering that global objects is an extensive topic, it would not be very productive to explain each object and their attributes separately. Instead, we will be creating a few projects to see firsthand how to work with different kinds of objects and the data types they return.

As we have already mentioned, an object, in combination with attributes, allows us to read the information from our admin and dynamically output it to create various functionalities. Let's start by familiarizing ourselves with the collection and product objects, which we will use to finalize the Custom collection project that we started in the previous chapter.

Custom collection

In the previous chapter, we created a `for` loop that outputs the names of the products from the outdoor collection, whose price is lower than $100.00. We did this by creating a `collection-form` snippet whose content we are outputting using the `for` parameter, combined with the `render` tag. We placed this at the bottom of our `collection.liquid` template file:

```
{% render "collection-form" for collections["outdoor"]
  .products as product %}
```

Inside our snippet, we added an `if` statement to check if the product price is higher than `10000`. If it is, we should output the product `form` tag and the product title:

```
{% if product.price > 10000 %}
  {% continue %}
{% else %}
  {% form "product", product, data-location: "collection" %}
    {{ product.title }}
  {% endform %}
{% endif %}
```

If we were to preview our code by vising the indoor collection page, we would only see a list of names, so let's try to improve this snippet by writing some code that will output the entire product block instead of only the product's name.

Let's start by refactoring the code inside the snippet by removing the `continue` and `capture` tags, refining our statement by removing the `else` statement, wrapping `{{ product.title }}`, positioning it above the product form, and finally removing the extra parameters from our product `form` tag:

```
{% if product.price < 10000 %}
  <p>{{ product.title }}</p>
  {% form "product", product %}
  {% endform %}
{% endif %}
```

The newly refactored code will do the same thing as it did previously, but now, it will be easier to understand and maintain.

Let's proceed by creating our product block. Currently, we only have a product name, so let's include the hyperlink that will redirect us to the actual product name when we click on the product's name.

We can do this by wrapping {{ product.title }} inside of the hyperlink tag, with its href attribute set to {{ product.url }}. This will return the relative path:

```
{% if product.price < 10000 %}
  <a href="{{ product.url }}">
    <p>{{ product.title }}</p>
  </a>
  {% form "product", product %}
  {% endform %}
{% endif %}
```

The addition of the hyperlink has ensured that we will be redirected to the actual product page by clicking the product's name. The next thing we need to do is include the image for each of our products.

We can do this by creating an image HTML tag inside the hyperlink tag, just above the product title, and setting its src attribute to {{ product | img_url }}. This will return a string to the location of the product image on Shopify's **Content Delivery Network (CDN)**:

```
{% if product.price < 10000 %}
  <a href="{{ product.url }}">
    <img src="{{ product | img_url }}"/>
    <p>{{ product.title }}</p>
  </a>
  {% form "product", product %}
  {% endform %}
{% endif %}
```

Our product block is starting to look a lot better, but we still need to show the price of our products. We can include the product price by wrapping `{{ product.price | money }}` inside the p HTML tags:

```
{% if product.price < 10000 %}
  <a href="{{ product.url }}">
    <img src="{{ product | img_url }}"/>
    <p>{{ product.title }}</p>
    <p>{{ product.price | money }}</p>
  </a>
  {% form "product", product %}
  {% endform %}
{% endif %}
```

> **Tip:**
> Notice that we have a small addition separated by a pipe character for both the image and price objects. This addition is called a **filter**, which helps us modify the output that we would otherwise receive from the object.

For example, if we were to call the product object with the price attribute, we would receive a number value without any format, such as 2599. However, by applying the money filter to the object, we have automatically changed the otherwise meaningless number into a string data type and formatted it according to our store-selected currency formatting, resulting in a $25.99 string.

We will not go into too much detail regarding filters right now, as we will be learning about them in the next chapter. For now, this basic information about filters will have to suffice. Let's return to our example.

So far, we have included the product name, image, and price, which check most of the necessary boxes that are needed to present a product. However, notice that the product form is currently empty. Let's change this by introducing an input element so that it's of the submit type, which should allow us to purchase the product directly from the collection page:

```
{% if product.price < 10000 %}
  <a href="{{ product.url }}">
    <img src="{{ product | img_url }}"/>
    <p>{{ product.title }}</p>
    <p>{{ product.price | money }}</p>
```

```
  </a>
  {% form "product", product %}
    <input type="submit" value="Add to Cart"/>
  {% endform %}
{% endif %}
```

With the addition of the **Add to Cart** button, we have created a button that should allow us to purchase the products directly from our collection page, without having to navigate to the product page. However, if we were to click it right now, we would encounter an error stating **Parameter Missing or Invalid: Required parameter missing or invalid: items**:

⚠ Something went wrong.

What happened?

Parameter Missing or Invalid: Required parameter missing or invalid: items

What can I do?

- Return to the previous page.

Figure 4.1 – Result of submitting a product form with missing parameters

The parameter that we are missing is the id property of the variant that we are looking to purchase.

Every product on Shopify can have up to three different sets of options. For example, a product can have multiple sizes, colors, and materials. Each combination of these three choices generates a unique number called a variant `id`, which tells our product `form` which combination of options it should place in the cart.

Note that these options are entirely optional as we can have a product without any variant options. However, even then, we still need to include the variant `id`. We should include this variant `id` as a `value` attribute of the `hidden` HTML input element with `id` as its `name` attribute:

```
{% if product.price < 10000 %}
  <a href="{{ product.url }}">
    <img src="{{ product | img_url }}"/>
    <p>{{ product.title }}</p>
    <p>{{ product.price | money }}</p>
  </a>
  {% form "product", product %}
    <input type="hidden" name="id" value="{{
      product.first_available_variant.id }}" />
    <input type="submit" value="Add to Cart"/>
  {% endform %}
{% endif %}
```

With the addition of the variant `id`, we now have a fully functional product `form`. By clicking the **Add to Cart** button, we can submit the request to the Shopify servers, where the product and its variant will be identified based on the submitted variant `id` and added to our cart.

Notice that we will also be automatically redirected to the cart page after clicking the **Add to Cart** button. This is the default behavior that we can rectify in one of two ways.

The first way would require us to add the `return_to` parameter to our product `form` tag, which will allow us to set the page we should return to after submitting the form. We can remind ourselves of how to use the `return_to` parameter by returning to the previous chapter and revisiting *The form tag* subsection, located under the *Theme tags* section. The other way would be to use the **Shopify Ajax API**, which we will learn about later in this book.

Let's return to the indoor collection for now. Let's look at the entire collection page and compare the initial collection products with our custom collection at the bottom. You will notice that besides being styled in a much better way than our collection, the initial collection products also contain a red sale badge, along with a regular and discounted price.

Each product in the admin contains two different fields for price, located under the **Pricing** section. The first field, named **Price**, allows us to set the current price of our product. However, the second field, named **Compare at price**, allows us to set the initial price to simulate a discount. We can access this field data by combining the `product` object with the `compare_at_price` attribute.

Let's return to our code on the collection page and modify it so that it includes both the sale badge and the compare price to match the initial collection. We can show the comparison price by wrapping `{{ product.compare_at_price | money }}` inside the `span` element and positioning it right after `{{ product.price | money }}`. Additionally, we can implement the sale badge by creating a simple string message inside the `span` tag and placing it under the `price` element:

```
{% if product.price < 10000 %}
  <a href="{{ product.url }}">
    <img src="{{ product | img_url }}"/>
    <p>{{ product.title }}</p>
    <p>{{ product.price | money }}<span>{{
        product.compare_at_price | money }}</span></p>
    <span>sale</span>
  </a>
  {% form "product", product %}
    <input type="hidden" name="id" value="{{
        product.first_available_variant.id }}" />
    <input type="submit" value="Add to Cart"/>
  {% endform %}
{% endif %}
```

We have included all the necessary elements for the `Custom collection` items. However, quite a few products do not have their comparison price and sale badges visible, which results from the **Compare at price** field being empty inside the product admin. We can resolve the second issue by modifying the `if` statement on our first line so that it shows products with a defined comparison price:

```
{% if product.compare_at_price != blank %}
  <a href="{{ product.url }}">
    <img src="{{ product | img_url }}"/>
    <p>{{ product.title }}</p>
    <p>{{ product.price | money }}<span>{{
        product.compare_at_price | money }}</span></p>
```

```
        <span>sale</span>
    </a>
    {% form "product", product %}
      <input type="hidden" name="id" value="{{
          product.first_available_variant.id }}" />
      <input type="submit" value="Add to Cart"/>
    {% endform %}
  {% endif %}
```

By modifying the `if` statement, we have ensured that we only display products on sale, consequently creating a custom on-sale addition for our initial collection. The only thing left to do is update the HTML formatting and add some CSS styling. While formatting and styling are entirely optional, we advise you to use the same formatting as it will be easier to follow up with future changes. The HTML formatted code, along with some basic CSS styling, can be found at the following GitHub link within the `Learning Projects` directory, under the name `Custom collection`: `https://github.com/PacktPublishing/Shopify-Theme-Customization-with-Liquid/tree/main/Chapter04/Learning%20Projects/Custom%20collection`.

This GitHub repository contains three files, each named according to the location where the code should be included.

If we preview our collection indoor collection now, we will see a significant improvement compared to when we started:

Figure 4.2 – Example of a complete custom collection project

So far, we have only mentioned objects whose attributes return a single value, such as `product.price`, `product.title`, or `product.first_available_variant.id`. However, they can also return an array or even act as helper tools for our programming logic.

One of the most used objects that returns an array is the `linklist` object. Combined with the `link` object, it will help us read the data from the menu within the navigation admin and help us create a custom navigation menu.

Custom navigation

For our next project, we will create a multi-level navigation menu specifically for our collection page. However, before we can learn about the `linklist` and `link` objects, we need to create a navigation menu. Let's get started:

1. We can create a new navigation menu by navigating to admin, clicking on **Online store** to expand it, and clicking on the **Navigation** link, where we can create the new menu by clicking the **Add menu** button:

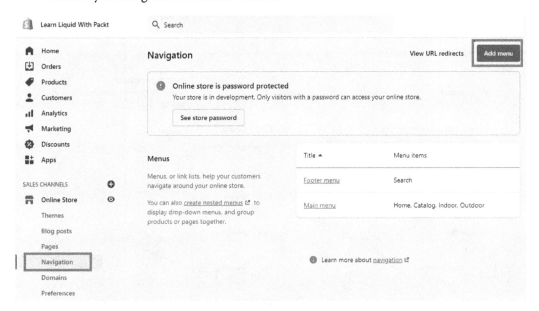

Figure 4.3 – Example of accessing the Navigation menu

2. After clicking the **Add menu** button, set the menu's title to **Indoor Navigation**. The page's handle will be automatically populated once we set the menu's name, so there is no need to modify it manually.

3.   We are now ready to populate the menu. Click on the **Add menu item** button and click on the **Link** field, which will automatically display a dropdown menu. For the first menu item, click on **Collections**, then **All collections**, which should immediately populate the menu item's name and link fields. Click on the **Add** button to finish adding the menu item.

4.   Repeat the previous step six more times by creating two menu items named and linked to the two collections **Indoor** and **Outdoor**, and then four menu items to any four products in our store.

5.   Once we have created the additional menu items, move the **Indoor** and **Outdoor** menu items under the **All collections** menu item to create a nesting menu. We can do this by clicking on the six dots before the menu item's name, and then holding and moving them over the **All collections** menu item until the indented blue line shows, at which point we should release the click:

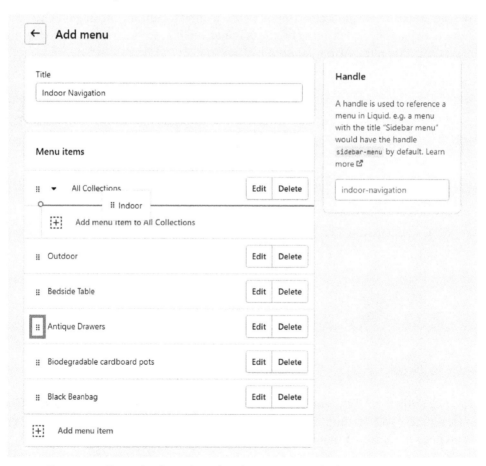

Figure 4.4 – Example of creating a dropdown menu inside the Navigation menu

6. Repeat the previous step and move the two product menu items under the **Indoor** collection menu item, which we have already moved under the **All collections** menu item. Repeat the same process for the **Outdoor** collection menu item and the two product menu items that are remaining.

If we have done everything right, we should end up with an **Indoor Navigation** menu containing a single menu item containing **All collections**, which contains two collection menu items containing two product menu items each. Suppose the menu format does not look the same after following the preceding steps. In that case, we can consult the Shopify documentation on nesting the menu items, where we can find more detailed instructions and a YouTube link on this topic.

For additional information on creating nesting navigation, please refer to `https://help.shopify.com/en/manual/online-store/menus-and-links/drop-down-menus`.

> **Important note:**
> The maximum amount of nesting menu items we can have inside a single menu is up to three levels long. We can consider the menu that we previously created as a three-level navigation menu, which is the limit.

Now that we have successfully created the navigation menu, we can start creating a variable that we will assign the value of the `linklist` object to, followed by the handle of the navigation we are trying to access. Remember that when accessing the object using the handle, we need to pluralize the object by adding the letter s at the end of the object:

```
{% assign collection-menu = linklists.indoor-navigation %}
```

With that, we have successfully saved the object of our indoor navigation to the `collection-menu` variable. Let's test it inside our `collection.template`. To do so, we will call the `collection-menu` object, followed by the `title` and `levels` attributes, just above `{% section 'collection-template' %}` so that we can see the name of our navigation and the number of nesting menus it has:

```
{% assign collection-menu = linklists.indoor-navigation %}
{{ collection-menu.title }} - {{ collection-menu.levels }}
```

If we preview our indoor collection, we will be able to see the name of our navigation and the number of nested levels the navigation menu has:

```
Indoor Navigation - 3
```

Now that we have confirmed that we have recovered the correct menu object, we can remove the `collection-menu.title` and `collection-menu.levels` lines. For the next step, we will need to use the `for` tag to loop over the array of links inside the object, which we can recover by calling the `collection-menu` object, followed by the `links` attribute:

```
{% assign collection-menu = linklists.indoor-navigation %}
  {% for link in collection-menu.links %}
    {{ link.title }}
  {% endfor %}
```

If we preview our **Indoor** collection now, we will notice that the only menu item that's visible on our page is the first level item, **All Collections**. Suppose we would like to loop over the additional nested menus located inside the **All Collections** menu item. In that case, we will need to create a second loop inside the first loop to recover the data from the second level.

The key difference is that we will not use `collection-menu` as our object. Instead, we will use the `link` object from our first `for` loop, combined with the `links` attribute, which will give us access to the array of links inside the `link` object:

```
{% assign collection-menu = linklists.indoor-navigation %}
{% for link in collection-menu.links %}
  {{ link.title }}
  {% for sub_link in link.links %}
    {{ sub_link.title }}
  {% endfor %}
{% endfor %}
```

By looping over the `link` object, we were able to recover the array links nested inside the **All Collections** menu item. Using the same technique, we can loop over the array of links inside the final level of the **Indoor** navigation menu:

```
{% assign collection-menu = linklists.indoor-navigation %}
{% for link in collection-menu.links %}
  {{ link.title }}
  {% for sub_link in link.links %}
    {{ sub_link.title }}
    {% for sub_sub_link in sub_link.links %}
      {{ sub_sub_link.title }}
```

```
    {% endfor %}
  {% endfor %}
{% endfor %}
```

We now have all the necessary elements to create a working on-hover navigation menu. The only thing left is to add some HTML tags and apply the necessary styling to create the on-hover dropdown effect. However, before we proceed with styling, let's try to be a bit more creative.

Notice that our menu links are just that – links. *But what if we wanted to be more creative by showing an image with every product menu item?* For this to work, we would need to identify which menu items are pointing toward the products, which we can do by using the type attribute and comparing whether the returned value is equal to the product_link string. Since we already know that only the third level of the navigation contains the product menu items, we will only include this feature inside the third for loop:

```
{% assign collection-menu = linklists.indoor-navigation %}
{% for link in collection-menu.links %}
  {{ link.title }}
  {% for sub_link in link.links %}
    {{ sub_link.title }}
    {% for sub_sub_link in sub_link.links %}
      {% if sub_sub_link.type == "product_link" %}
        <img src="{{ sub_sub_link | img_url: "250x250"}}"/>
      {% endif %}
      {{ sub_sub_link.title }}
    {% endfor %}
  {% endfor %}
{% endfor %}
```

Now, it makes sense that this should work and that, as a result, we should be able to see four images of our products inside the navigation menu. However, note that sub_sub_link is still only a link object, and the link object does not have an image attached to it. To show the image that's attached to the product, we will have to recover the object of the product that our link is pointing to.

We can easily do this by modifying the `sub_sub_link` object inside the IMG tag so that it includes the `object` attribute:

```liquid
{% assign collection-menu = linklists.indoor-navigation %}
{% for link in collection-menu.links %}
  {{ link.title }}
  {% for sub_link in link.links %}
    {{ sub_link.title }}
    {% for sub_sub_link in sub_link.links %}
      {% if sub_sub_link.type == "product_link" %}
        <img src="{{ sub_sub_link.object | img_url:
              "250x250" }}"/>
      {% endif %}
      {{ sub_sub_link.title }}
    {% endfor %}
  {% endfor %}
{% endfor %}
```

With the addition of the `object` attribute, we now have access to the entire product object, including its title, price, images, and all other data. In comparison, `sub_sub_link.object.price` would return precisely the same results as writing `product.price` would, which we used in the previous project to display the prices of our `Custom collection` products.

The only thing left to do now is provide our code with some HTML format and apply the necessary styling. The HTML formatted code, along with some basic CSS styling, can be found at the following GitHub link, under the name `Custom navigation`:

`https://github.com/PacktPublishing/Shopify-Theme-Customization-with-Liquid/tree/main/Chapter04/Learning%20Projects/Custom%20Navigation`.

This GitHub repository contains two files, each named according to the location where the code should be included.

While working on this project, we were able to create a fundamental version of the mega menu, allowing us to output any image attached to our pages easily. While it might not look impressive, the knowledge we have learned from this project has taught us how to create any navigation menu, as well as a custom subcollections page where we can output other collections.

So far, all our projects have been related to recovering the predefined data from our admin. *However, what if we needed to capture data regarding the customization of a particular product, show the choice on the cart page, and then forward the captured data with the order?*

Product customization

For our next project, we will be creating a custom HTML input on the product page, which will allow us to capture any data that the customer may input and learn how to forward the value of the input, along with the order itself.

To achieve this functionality, we will be using the `line_item` object. `line_item` represents each item within our cart. We can access the `line_item` object through the `cart` object, followed by the `items` attribute, which will provide us with access to the `line_item` object for each product.

Before we can use `line_item object` to output the data on our cart page, we will need to create a field to capture this data. Let's start by navigating to our `product.liquid` template and find the opening tag of the `product` form. As we can see, the `product` form tag is not present in this file, but we do have a `section` tag, which, as you may recall from the previous chapter, allows us to render a static section.

Let's proceed by navigating to the `product-template.liquid` section file. We can access it by hovering over the section tag's name and clicking on the small arrow. Alternatively, we can find it inside the `Sections` directory. After finding the `product` form tag, we can start creating the customization feature by adding an HTML input tag of the `text` type, which we will be using to capture information related to each specific product. We can add this field at the top of the `product` form tag, just above the first `unless` statement:

```
{% form 'product', product, class:form_classes, novalidate:
  'novalidate', data-product-form: '' %}
            {{ form | payment_terms }}
      <input type="text" placeholder="Your Name"/>
      {% unless product.has_only_default_variant %}
```

If we try to preview our product page, we will see an input field with **Your Name** as the placeholder:

Figure 4.5 – Example of a custom field on the product page

However, if we fill in the input, add the product to the cart, and visit our cart page by clicking the cart icon in the top-right corner, we will notice that we did not capture the data successfully with the product. To save the line_item data value, we will need to modify the HTML input by adding the name property in name="properties[Your Name]" format:

```
<input type="text" name="properties[Your Name]"
placeholder="Your Name"/>
```

The name property is a predefined attribute that allows us to capture the input element's value, followed by a mandatory keyword, called properties, and a pair of square brackets. Any value inside these square brackets – in our case, **Your Name** – will serve as that property name.

With the name attribute in place, if we fill in the input field, click on the **Add to Cart** button, and visit the cart page, we will notice that both the line_item attribute's name and value were successfully captured and chained to each product.

When dealing with newer themes, such as the one we are using, this would be the last step in capturing customization for each product. However, for some older themes, which are still quite present today, we would need to write some code that will display the line_item properties inside the cart page.

Before writing this code, we will need to identify where we should add our code. As we mentioned previously, to access the line_item object, we will need to use cart.items to recover an array of products in our cart, which should already be present in our cart.liquid section file:

```
{% for item in cart.items %}
{% endfor %}
```

With the cart.items loop, we have gained access to the objects of each product, which, similar to the product object, provides us with access to various attributes, such as the product object's title, price, and the quantity that we have added to the cart, as well as its properties.

We can access the properties of each product by using item as our object, followed by the properties attribute. However, since we can capture multiple properties for each product, item.properties will return an array of data, meaning that we will need to use the for tag:

```
{% for property in item.properties %}
   {{ property }}
{% endfor %}
```

Note that {{ property }} is considered an array data type since it contains both the name of the line_item property and its value. If we were to use this code to output the line_item code on our page, we would end up with both values stuck together. Since using another for tag to go over the array that contains two elements would be excessive, we can use the first and last filters to output the split values:

```
{% for property in item.properties %}
  <span>{{ property.first }}</span>:<span>{{ property.last
    }}</span>
{% endfor %}
```

The `first` and `last` filters, as their names suggest, allow us to access the first and last element inside an array. However, since we are looking to split our array into two separate elements, it is perfect as it helps us avoid writing another `for` loop. We will learn more about this and other filters in the next chapter.

> **Important note:**
> Adding the same product variant multiple times to the cart with a different `line_item` property value will not override the product we previously added to the cart, nor affect its `line_item` property value. Instead, we will end up with each product positioned on a new line, as if it were a different product.

The only time when we can include multiple variants of the same product on the same line is if the product variant contains the same data, the `line_item` property included. Here is an example:

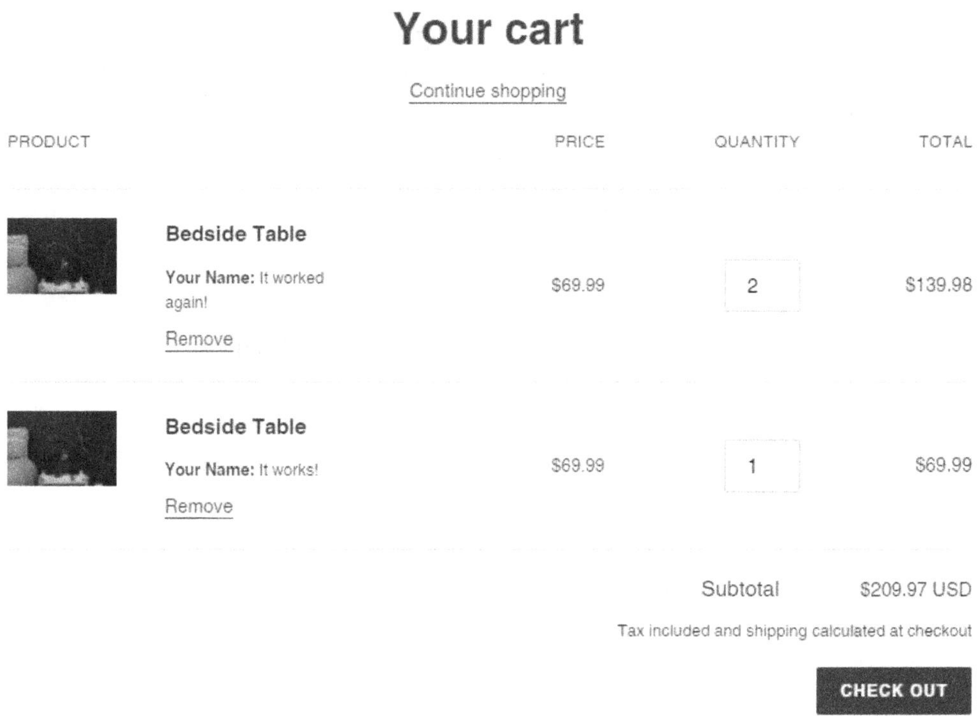

Your cart

Continue shopping

PRODUCT		PRICE	QUANTITY	TOTAL
	Bedside Table **Your Name:** It worked again! Remove	$69.99	2	$139.98
	Bedside Table **Your Name:** It works! Remove	$69.99	1	$69.99

Subtotal $209.97 USD

Tax included and shipping calculated at checkout

CHECK OUT

Figure 4.6 – Example of the line item property chained to different product variants

If we were to click the **CHECK OUT** button, we would notice that the same line_item values were visible inside the checkout summary as well. Note that the same line_item property will be visible inside the order admin if we complete the purchase through our checkout page.

> **Important note:**
>
> We can only capture the line_item properties through Shopify's default checkout. If we were to complete the payment through any other checkout, such as **PayPal**, we would not be able to attach the line_item properties to our products, and they will not be visible inside the order admin.

With that, we have learned how to capture custom data for each of our products and display them on the cart and checkout pages. However, sometimes, we will be forced to recover the line_item data and visually hide it from both the cart and checkout pages. We can easily hide the line_item data from the cart page with some CSS code, *but how are we supposed to modify the code on the checkout page, considering that we do not have access to it?*

This is where the underscore character will help us. If we are looking to collect line_items and have them visible when we receive the order in our admin, while at the same time visually hiding the line_items properties from the checkout page, we would need to modify the name property inside the product.liquid section file to include the underscore inside the square bracket:

```
<input type="text" name="properties[_Your Name]"
placeholder="Your Name"/>
```

Any property name with the underscore as the first character inside the square bracket will not be visible on the checkout page. However, it will still be visible once we receive the order in our admin.

Besides allowing us to hide the line_items properties from the checkout page automatically, the underscore character also helps us create a more automatic process for hiding the line_items property from the cart page, without having to write CSS code for every line_items property that we want to hide.

With the introduction of the underscore character, we now have one uniquely distinctive character that we can use to filter the line_items properties we would like to show and which ones we would like to hide.

Since we are dealing with string-type data, we can use the `truncate` filter, which, as its name suggests, allows us to truncate a string. The `truncate` filter accepts the following two parameters:

- The first parameter is a required number value, which allows us to set how many characters we expect the `truncate` filter to return.

- The second parameter is an optional parameter, which allows us to set a specific string value we would like to append to each returned string value. Note that if we do not include the second parameter, by default, the `truncate` filter will append three dots to the end of the string, which will count inside the previous parameter number value's `count`.

Since we are looking to check whether the `line_items` key value contains the underscore character as the first character in the string, we can apply the `truncate` filter and set the first parameter to `1`. However, we will also need to include the second parameter and set it equal to an empty string to avoid the previously mentioned ellipsis:

```
{% for property in item.properties %}
  {% assign first_character_in_key = p.first | truncate: 1,
    '' %}
  {% unless first_character_in_key == '_' %}
    <span>{{ property.first }}</span>:<span>{{
      property.last }}</span>
  {% endunless %}
{% endfor %}
```

Note that throughout this project, we have only used a single `text` input type. However, we are free to use any input type at our disposal, including the `date`, `color`, `radio`, and `select` inputs. The only limitation is that we cannot use the `file` upload input type with Ajax. The product `form` tag will also need to include the `enctype="multipart/form-data"` attribute to capture the file upload input value.

With this project, we have learned how to create a valuable feature that will allow us to create specific customization for each product template, or even each product separately. We can design it as a simple feature with a single input or create an entire form with various inputs to fill in before purchasing a product.

We can also use the **Shopify UI Elements** generator to create any number of `line_items` easily and simply paste them inside our product templates. We can find the Shopify UI Elements generator at `https://ui-elements-generator.myshopify.com/pages/line-item-property`.

Let's say that we were looking to save page-specific data for some of the pages in our admin. However, if we were to navigate to our admin and open a product, collection, or any other page, we would notice that each page contains a predefined number of fields to store data. This is where the metafields objects come to the rescue.

Improving the workflow with metafields

Metafields are global objects that allow us to store additional data inside our admin and output it to the storefront. As such, they are both powerful and necessary tools for creating complex designs with unique content.

Metafields consist of three mandatory elements:

- The **namespace**, which we can use to group different metafields, so long as they share the same namespace.
- The `key` attribute, which allows us to access a specific metafield.
- A **value** where we can store an integer, string, or `json_string` data type.

We can also use a `description` field for the short description of a metafield, which is optional compared to the previous three.

Metafields, while powerful tools, had quite a disadvantage as they were only available through third-party apps. However, since the Shopify Unite 2021 event, the metafields functionality has become available natively inside the Shopify dashboard, and not only that, but it has received quite the upgrade!

The `metafields` functionality, while functional, at the time of writing, is only partially available as we can only access the product and variant metafields. The `page`, `blog`, `article`, `collection`, `order`, `customer`, and `shop` metafields are pending to be released. For this reason, we will learn how to handle these metafields through a third-party app and the Shopify dashboard.

Considering that most of today's themes still rely on third-party apps for their metafields needs, let's start by learning how to utilize a metafields app to create custom content for our pages.

Setting up a metafields app

For us to use and access metafields objects, we will need to either install an app or a browser extension that allows us to use this functionality. For this book's purposes, we will proceed by installing an app so that we can familiarize ourselves with that Shopify app market as well:

1. To install an app on our store, we need to navigate to our admin and click on **Apps**, just above **Sales Channels**. Inside the **Apps** section, we will be able to see a list of all the apps that we have installed on our store. However, as we currently do not have any, the list is empty. Let's change this by clicking on the **Shop for apps** button in the top-right corner:

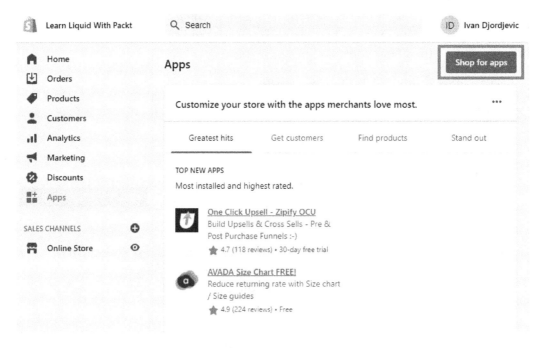

Figure 4.7 – Example of starting the app install process

2. Clicking the **Shop for apps** button will open the Shopify app store inside a new window. Here, we can browse trending apps and search categories for various apps. However, since we know what we are looking for, we will search for the metafields app by typing the keyword `metafields` inside the search field.

3.   After submitting a search, we will receive a significant number of apps that we can use, and most of them have a free plan available. For our purposes, we will be using the **Metafields Guru** app. While which app you choose is entirely optional, since they all offer the same functionality, we advise you to use the same app here as it will be easier to follow up with future development. Upon clicking on the app, you will be redirected to the app window.

4.   After opening the app's landing page, we will be able to see more information about the page, which we should always read to check whether the app offers the features that we need. Let's click on the **Add app** button, which should immediately redirect us to our store and start the app installation process. On the other hand, if we are not, we will need to log in by submitting a store URL inside the popup window and clicking on the **Log in** button, which will start the app installation process:

Figure 4.8 – Example of the login popup on the Shopify app store

5.   After starting the installation process, we will see one last window, where we will need to provide the app with requested access to our store. Since we are working on a development store, we can immediately proceed by clicking on the **Install app** button. However, if we are working on a managed store for a client, it is strongly advised that we never install the app on someone else's store, even if they have asked us to do so. The store owner should be the one to install any necessary app once they know about all the personal information that the app will be collecting. Otherwise, we might be accountable for any possible problems.

With the **Metafields Guru** app now installed, we will see the first view within the app, where we can see all the different types of pages we can create metafields for:

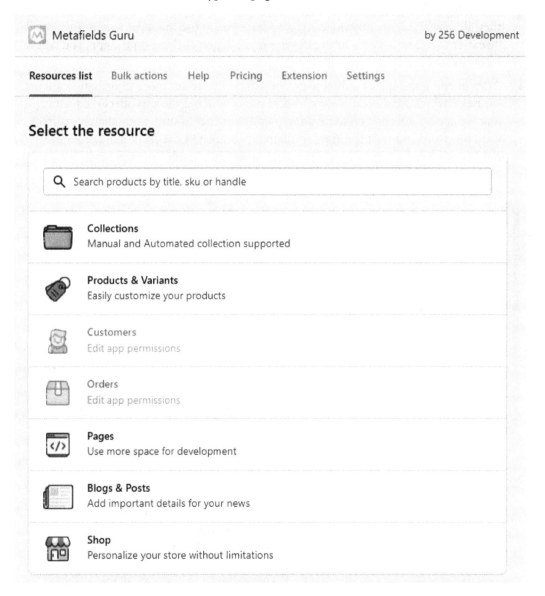

Figure 4.9 – Different types of pages we can use metafields objects for

As shown in the preceding screenshot, metafields are powerful tools that allow us to customize any section of our store. So, let's begin by creating the metafield for a product page:

1. The first thing we need to do is navigate to the **Product** section of our admin, select any product of our choosing, and click on it.

2. After opening the product page, click on the **More actions** button in the top-right corner to reveal the **Edit Metafields** button. Click on it to launch the **Metafields Guru** app:

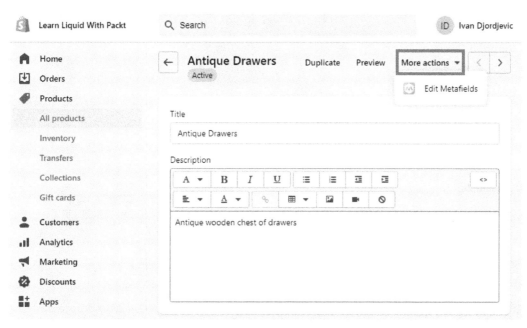

Figure 4.10 – Example of starting the Metafield Guru app for a specific page

3. Once inside the app, we will be able to see a list of all the metafields that we have for this specific product, or even the variant if we select the **Variants** card. However, since we don't have any, the screen is empty. Let's create our first metafield by clicking the **Create metafield** button.

4. Inside this first field, we can select the data type that we will save inside this specific metafield. This is where we will be populating the other three fields with the key, namespace, and metafield values. We can leave the String value selected for our example and then, inside the key field, type in the word example.

The namespace field already contains the global keyword, but we can enter any type of text in the final field. For example, we will use Metafields are awesome!. After filling in all the fields, save your changes by clicking the **Save** button in the top-right corner:

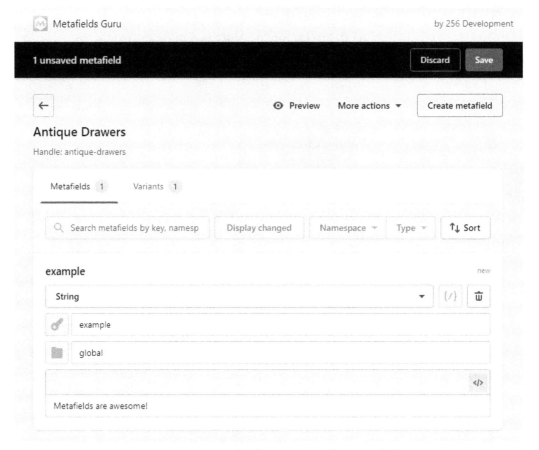

Figure 4.11 – Example of creating a product metafield

Now that we have learned how to create a metafield, it is time to learn how to output the previously saved data to our storefront.

Rendering the metafields value

We can access the metafields object through the object of the page we created the metafield for, followed by the metafields object, followed by namespace, followed by key:

```
{{ product.metafields.global.example }}
```

We can enter the metafield code anywhere where we have access to the `product` object. However, for our purposes, we should place the code inside the `product.liquid` section, just above the `line_item` input that we included in one of our previous projects.

Now that we have created the product metafield and have the metafield object code in place, all we need to do is test it out. However, remember that we have created a metafield for a specific product only, meaning that we can only preview it on that specific product. If we were to preview the product we created the metafield for, we would see the following metafield string value displayed correctly on our page:

Metafields are awesome!

With that, we have successfully created and displayed a single metafield value. However, *what if we had a larger number of similar metafields? How would we output all of them?* We initially mentioned that we could use the `namespace` attribute to different group metafields together, so long as they share the same `namespace`. Using `namespace` with the `for` tag, we can loop over all the metafields with the same `namespace` and recover their values:

```
{% for value in product.metafields.global %}
   {{ value }}
{% endfor %}
```

Running the previous code would recover any metafields with the `global` keyword as its namespace. However, since we are not using the `key` attribute this time, we will receive the results in array format:

exampleMetafields are awesome!

To split the array, we will need to use the `first` and `last` filters, as we did previously, to split the `line_item` object on the cart page:

```
{% for value in product.metafields.global %}
   {{ value | first }}:{{ value | last }}
{% endfor %}
```

Using the `first` and `last` filters, we have successfully split the array into two separate elements that we can now use in any way we require.

With that, we have taken a significant step forward by learning how to output single and multiple metafield values through a third-party app.

While helpful, adding metafields through a third-party app requires quite a bit of hardcoding, which we can avoid by creating metafields through the Shopify dashboard instead. However, while metafields have received a powerful upgrade, they also have a prerequisite, which we learned about previously. For this reason, we will learn how to set up and handle metafields through the Shopify dashboard in the following chapters to gain a better understanding of this.

Using metafields, we can now add well-organized, unique content to each page or even create complex functionalities. For example, with metafields, we can create product accordions, show hand-picked recommended products, and use them to show the expected delivery time for each product/variant and other features. We are only limited by our inspiration.

The few projects that we have covered by now might look inadequate compared to the number of different types of global objects we have at our disposal. However, the knowledge we have attained through these projects has set us on the right path of understanding objects entirely. Our primary focus was not simply listing objects and their attributes, which we can quickly get from Shopify documentation, but learning how to use different objects and attributes through real-life projects that we will be working on tomorrow as Shopify experts.

Before we can say that we understand how objects work, we will need to learn more about content objects, without which we would not be able to output any content on our pages.

Content and special objects

Previously, we learned how to use global objects to output the data from our admin on our templates and section files. Now, it is time to learn how to use content objects to output the content of template and sections files, as well as any other assets that are necessary for the Shopify store to operate. We can divide the content objects into three separate groups: `content_for_header`, `content_for_index`, and `content_for_layout`.

All three types of content objects have descriptive names that tell us what they do. However, to ensure that we fully understand their importance, we will provide a small explanation for each.

The content_for_header object

`content_for_header` is a mandatory object located inside the `theme.liquid` layout file – more specifically, inside the HTML `<head>` tag:

```
{{ content_for_header }}
```

The whole purpose of this object is to load all the scripts that Shopify requires into the document's header dynamically. Among these scripts, we can find Shopify and Google Analytics scripts, and even the scripts required for some Shopify apps.

The content_for_index object

`content_for_index` is a not mandatory object and it's located within the `index.liquid` template, which we can find inside the `Template` directory:

```
{{ content_for_index }}
```

However, this object allows us to dynamically output dynamic sections from the theme editor, making it mandatory. Without it, we would not be able to output any content from our theme editor.

The content_for_layout object

`content_for_layout` is the final mandatory object located inside the `theme.liquid` layout file. This object allows us to load content that's been dynamically generated by other templates, such as `index.liquid` and `collection.liquid`:

```
{{ content_for_layout }}
```

Note that it is not possible to delete the `content_for_header` and `content_for_layout` objects from their respective positions, so we don't need to worry too much about them. However, it is essential to know what each of the content objects does since even though we cannot delete them, we can comment them out, which will create issues with our storefront.

Finally, besides global and content objects, we also have another set of objects that we can only use under specific circumstances. The only two objects we currently have at our disposal are the `additional_checkout_buttons` and `content_for_additional_checkout_buttons` objects, which will provide us with a way to dynamically generate a set of buttons leading to a third-party provider's checkout page.

However, note that deciding which buttons will be visible depends on the payment methods that we have set in the admin, as well as some other parameters. For example, for the **Apple Pay** checkout button to be visible, besides enabling the payment method in the admin, the customer will also have to use an Apple device to see it.

The `additional_checkout_buttons` object allows us to check whether the store offers payment using third-party payment providers, such as PayPal, Apple, and others. Suppose we have enabled a payment gateway to some third-party payment providers. The `additional_checkout_buttons` object, in combination with an `if` statement, will return `true`, which will allow us to use our next special object; that is, `content_for_additional_checkout_buttons`.

After confirming that we have enabled these payment methods with third parties, we can use the `content_for_additional_checkout_buttons` object to generate the buttons for those payment providers:

```
{% if additional_checkout_buttons %}
  {{ content_for_additional_checkout_buttons }}
{% endif %}
```

While the placement of the code is entirely optional, it is usually placed on the cart page, next to the default **Checkout** button.

Summary

In this chapter, we learned about different types of objects while considering various projects. Within our first project, `Custom collection`, we learned how to access single-valued attributes within the `product` object by creating a custom collection feature with a fully functional product form. The second project, `Custom navigation`, taught us how to access and handle objects whose attributes return an array. Working through the `Product Customization` project, we understood how to capture multiple data types, chain them with the selected variant, and output the same data on both the cart and checkout pages.

Besides working on different projects, we also had the pleasure of learning how to install new apps from the store and how to use a third-party app to create additional input fields within our admin, which gave us access to `metafields` objects.

Finally, we learned about the different types of content and special objects, why some are mandatory, where we can find them, and how we can use some of them to connect our store to various payment providers outside Shopify.

The knowledge that we have attained through this chapter will be especially helpful in the next chapter, where we will learn more about the filters that we have been referencing through our projects.

Questions

1. What are we missing in the following block of code to make `form` functional?

```
{% form "product", product %}
  <input type="hidden" value="{{
  product.first_available_variant.id }}" />
    <input type="submit" value="Add to Cart"/>
{% endform %}
```

2. How can we get access to the `product` object through a link defined in the admin navigation?

3. What are the two approaches to accessing single and multiple `metafield` objects?

4. What adjustment do we need to make to the input element if we were looking to capture the `line_item` value and hide it on the checkout page?

```
<input type="text" name="properties[Your Name]"
placeholder="Your Name"/>
```

Practice makes perfect

In the previous chapter, we worked together through various projects and implementations. However, we can only gain a proper understanding by working on the projects ourselves and forcing ourselves to make that next step. So far, we haven't completed any personal projects as we were still learning the basics. However, with objects behind us, we are now set to start creating our own solutions.

The few mini-projects we will cover here will help us solidify some of our already attained knowledge from the previous chapters. It will also push our limits by forcing us to think outside of the box and find solutions for the problems that we haven't faced so far.

Each project will contain detailed information about what we need to do to help us achieve the results.

We recommend working on each project independently from the previous chapters since this will ensure that we have truly understood what we have learned so far.

No single project has either a correct or incorrect solution. However, if by any chance we get stuck, we can always consult the project solution at the end of this book.

Project 1

For our first project, we will be creating a custom collection on a general page. However, the difference between this project and the one we worked on previously is that this custom collection will be a dynamic and reusable code block. Depending on the page name, we should see different products in the featured collection.

Here are the steps for the assets:

1.   Create a new page template called `featured-collection.liquid`.

2.   Create a new page, name it similarly to one of the collection pages on our store, and assign it the new page template that we created previously.

3.   Create the new menu item inside the current main menu navigation, called `Projects`, and add the newly created page as a nested menu inside the `Projects` menu item.

4.   Create a new snippet file named `custom-collection`.

Here are the assignment steps:

1.   Include the `custom-collection` snippet inside `featured-collection.liquid` with `collection` as a parameter. Since the page shares a name that's similar to our collections', we should use the page handle to create the `collection` object.

2.   Using the `collection` object that we passed previously, create a custom collection using the `for` tag to display no more than four products.

3.   The products should contain an image, title, vendor, regular price, comparing price visible, and a working product form.

4.   If a product has more than one variant, include a dropdown so that we can select the exact variant we are looking to select.

5.   After submitting the form, return us to the same page we were previously on.

6.   Once complete, any page with `featured-collection.liquid` assigned as its template should display different types of products based on the page name; for example, Indoor or Outdoor.

Project 2

For our second project, we will be creating a subcollection template page where we will be able to output different collection pages. The code that we will create, similar to the previous project, should be reusable. We should receive different results based on the name of the page we have assigned the template to, and the collections that have been assigned to the navigation menu with the same name.

Here are the steps for the assets:

1. Create a new page template called `page-subcollection.liquid`.

2. Create a new page and assign it to the new page template that we created previously.

3. Create the new navigation menu and give it the same name we gave the page that we created previously.

4. Inside the new navigation menu, include no less than six collection menu items.

5. Create a new snippet file called `custom-subcollection`.

Here are the assignment steps:

1. Include the `custom-subcollection` snippet inside `page-subcollection.liquid` with `subcollection` as a parameter. Since the page shares a name that's similar to our navigation menu, we should use the page handle to create the `subcollection` object.

2. Using the `subcollection` object that we passed previously, use a `for` tag to create a list of all the collection pages inside the specific navigation menu.

3. The collection list should contain an image and a title.

4. If the collection does not have the image assigned to it, we should take the image of the first product inside that collection and show it as the collection image.

5. Once complete, any page with `custom-subcollection.liquid` assigned as its template should display different results based on its name, as well as the collection that was assigned to the navigation menu with the same name.

5
Diving into Liquid Core with Filters

In the previous two chapters, we learned about all the different Liquid tags and objects, and now, we will be focusing on the last of the Liquid Core features, which is **filters**. We have referenced filters on a few occasions now, but *what exactly is the purpose of filters?* Filters are methods denoted by a pipe character, |, through which we can manipulate different data types, including strings, numbers, variables, or even objects, making it a compelling feature.

We can split the chapter into the following topics:

- HTML and URL filters
- Enhancing the product media gallery
- Building product accordions
- Math and money filters
- Exploring the additional filters

By the time we complete this chapter, we will precisely understand how much power filters provide us. Similarly, as with the previous chapter, instead of simply listing and going through all of the filters, we will only explain some essential filters through a series of small projects.

First, we will learn about generating HTML elements through **HTML and URL filters**. Second, understanding **media filters** will help us with the product media gallery, one of Shopify's latest additions and one of the most sought-after features. Third, by working on the product accordions project, we will learn how to manipulate the string and array types of data through their respective filters. Lastly, working on a product price discount project will provide us with the necessary knowledge of **math and money filters**.

Technical requirements

While we will explain each topic and present it with accompanying screenshots, you will need an internet connection to follow the steps outlined in this chapter, considering that Shopify is a hosted service.

The code for this chapter is available on GitHub: `https://github.com/ PacktPublishing/Shopify-Theme-Customization-with-Liquid/tree/ main/Chapter05`.

The Code in Action video for the chapter can be found here: `https://bit.ly/3zmum4j`

Working with HTML and URL filters

In the previous chapter, we had the chance to see a type of URL filter when we worked on outputting the product images, `{{ image | img_url: "400x400" }}`. *However, what exactly are URL filters?*

URL filters are methods that allow us to output a direct path to the assets on Shopify's **Content Delivery Network** (**CDN**). We can use URL filters in many ways. However, URL filters alone only provide a string path to the requested asset. Therefore, to find them helpful, we must pair them with HTML tags, such as the `image` tag, inside which we can add the string path to a specific asset as the `href` attribute. Alternatively, we can combine URL filters with **HTML filters** to automatically generate the necessary HTML element with the necessary attributes. Let's see them in action now.

In *Chapter 1, Getting Started with Shopify*, we learned about the `Assets` directory inside our theme files and how it contains all of the internal assets that our theme requires, such as stylesheets, JavaScript files, font files, and images. However, we will first need to load these files within our theme, by following the next steps, since they will not automatically be accessible to us by simply uploading them within the `Assets` directory:

1. We can retrieve the path to a file within the `Assets` directory by encapsulating the file's name within quotation marks, followed by a pipeline, followed by the `asset_url` filter:

    ```
    {{ "theme.css" | asset_url }}
    ```

2. In the previous example, we have used the name of the stylesheet file within our store in combination with `asset_url`, which will provide us with the string path toward this specific file: `https://cdn.shopify.com/s/files/1/0559/0089/7434/t/4/assets/theme.css?v=10188701410004355449`

3. Now that we have recovered the path toward our location, as previously mentioned, we have two choices. The first option is to use an HTML `link` tag to link the CSS file with our theme:

    ```
    <link rel="stylesheet" href="{{ "theme.css" | asset_url }}">
    ```

    Our theme will now have full access to the `theme.css` file and its rules with the `link` tag in place:

    ```
    <link rel="stylesheet" href="//cdn.shopify.com/s/files/1/0559/0089/7434/t/4/assets/theme.css?v=10188701410004355449">
    ```

4. However, as we have previously mentioned, besides using the HTML `link` tag, we can also combine URL filters with an HTML filter to generate the necessary HTML attributes:

    ```
    {{ "theme.css" | asset_url | stylesheet_tag }}
    ```

In the previous example, we used the asset name whose path we are looking to recover, followed by the `asset_url` filter, which would usually only return the string path. However, pairing it with `stylesheet_tag` will automatically generate the HTML `link` tag with all the necessary attributes:

```
<link href="//cdn.shopify.com/s/files/1/0559/0089/7434/t/4/assets/theme.css?v=10188701410004355449" rel="stylesheet" type="text/css" media="all">
```

Besides being a lot cleaner, the main difference between the two approaches is that `stylesheet_tag` does not accept additional parameters. So, for example, if we were looking to change the `rel` attribute to preload, modify the `media` attribute, or even include the `defer` attribute, we would have to use the first approach and include the asset file using the HTML `link` tag.

We have now learned how to connect our theme file with the necessary stylesheet. However, note that we will be using a different type of HTML filter depending on which file we are looking to access.

For example, if we were looking to output the content of the `theme.js` file into our theme, we would use a similar approach, the theme filename followed by `asset_url` to get its path, but instead of using `stylesheet_tag`, we will use `script_tag`:

```
{{ "theme.js" | asset_url | script_tag }}
```

Using `script_tag`, we will automatically generate and include the HTML `script` tag in our theme. However, do note that similarly as with `stylesheet_tag`, `script_tag` also does not accept any parameters:

```
<script src="//cdn.shopify.com/s/files/1/0559/0089/7434
/t/4/assets/theme.js?v=2017768116492187958" type="text/
javascript"></script>
```

Besides `stylesheet_tag` and `script_tag`, we also have access to `img_tag`. Accessing an image file within the assets can be done using the name of the image file, followed by `asset_url`, followed by `img_tag`:

```
{{ "ajax-loader.gif" | asset_url | img_tag }}
```

The crucial difference between `img_tag` and the previous two filters is that `img_tag` does accept additional parameters.

In *Chapter 4, Diving into Liquid Core with Objects*, we had the chance to see the `img_url` filter in action by using it to return the product image URL string, which we combined with the HTML `img` tag to output the product image to our storefront:

```
<img src="{{ product | img_url: "300x300" }}"/>
```

Notice that besides the `img_url` filter, we have also used the `size` parameter to set the limit size of our image, which is one of the three parameters that we can use with the `img_url` and `img_tag` filters. However, note that both types of filters use different parameters, which we will explain shortly through the following mini project.

Building a product gallery

Inside this mini project, we will learn how to output the necessary elements for the product gallery:

1. Let's start by creating a new page named Product Gallery and a new page template named product-gallery, which we will assign to the previously created page.

2. Once we have created the page and assigned the appropriate template, we should identify one product with more than one image and recover its handle. We will use gardening gloves for our selection, which is one of the products that we previously imported from the product-data.csv file in *Chapter 3, Diving into Liquid Core with Tags*. We can retrieve the product handle by previewing the product page and copying the page handle from the page URL. Alternatively, we can retrieve it by navigating to the product page inside the admin section and copying the page handle from the page URL link at the bottom of the page.

3. Now that we have recovered the product handle, let's start by creating a product object through its handle and assigning it to a variable. To achieve this, we can use the previously learned method of accessing the page object through its handle. However, compared to the previous chapters, where we learned how to access the product object through its collection, accessing the product object directly using its handle is slightly different.

    Instead of pluralizing the product object name, we will be using a global object tag named all_products, which gives us access to all products in our store.

    > **Important note:**
    > While all_products is quite a practical method that allows us to access any product directly through its handle, it comes with a limitation that we can only run it 20 times per page. This means that if we need to recover more than 20 specific products on a single page, we will need to recover them by looping over a collection.

4. We can access the product object through its handle by using the all_products global object, followed by the handle of the product we are trying to access. For our example, we will be using square brackets, []. However, we could have also used the dot (.) annotation:

```
{% assign product_object = all_products["gardening-
    gloves"] %}
```

5. With this, we now have access to the gardening gloves product object. To access all images attached to a product, we will use the `product_object` variable as our object, followed by the `images` attribute, to recover the array of images for the specific product. Since we are dealing with an array, we will have to use the `for` tag to loop over them:

```
{% assign product_object = all_products["gardening-
    gloves"] %}
{% for image_item in product_object.images %}
    <img src="{{ image_item | img_url }}"/>
{% endfor %}
```

We have now successfully extracted an array of product images, which we can use to create powerful galleries using various slider plugins. However, notice that the size of the images we have on our storefront is relatively small.

This is because Liquid always defaults to a size of `100x100` if we do not introduce the `size` parameter. Let's introduce the `size` parameter by limiting the image dimension to 300 px in width and 300 px in height:

```
{% assign product_object = all_products["gardening-gloves"] %}
{% for image_item in product_object.images %}
    <img src="{{ image_item | img_url: "300x300" }}"/>
{% endfor %}
```

Notice that our images now have 300 px in width but only 200 px in height even though we have specified that we would like 300 px in height. This is because the `size` parameter can only limit the image size by decreasing its size to match the assigned values. It cannot change the image aspect, nor can it increase the size of the image beyond the original image size.

In the previous example, we have used `"300x300"` to limit our product images in both width and height. However, we can also limit only one side by using `"300x"` to set the image width to 300 px or `"x300"` to limit the height to 300 px. If we only specify one of the two values, Shopify will automatically calculate the dimension of the image while maintaining the image aspect.

The second parameter we can use with the `img_url` filter is the `crop` parameter, which allows us to crop the image to the specified size when combined with the `size` parameter. The `crop` parameter can have five different values:

- `top`
- `center`
- `bottom`
- `left`
- `right`

Using the correct option, we can specify which side of the image we would like to crop out. For our example, we can use the `center` option to ensure that the image is cropped equally from each side:

```
{% assign product_object = all_products["gardening-gloves"] %}
{% for image_item in product_object.images %}
  <img src="{{ image_item | img_url: "300x300", crop:
    "center" }}"/>
{% endfor %}
```

By changing the image size using the `crop` parameter, we have also changed the image's aspect, as all the images are now exactly 300 px in width and 300 px in height.

The last two parameters that we can use with the `img_url` filter are `scale`, which allows us to specify the pixel density of the image using the 2 and 3 as its option values, and `format`, which is a quite interesting parameter that allows us to specify the format of the displayed image. The two acceptable values for the `format` parameter are `jpg` and `pjpg`.

Using `pjpg`, we can convert the image format to **Progressive JPEG**, automatically loading a full-size image and increasing its quality gradually instead of loading the image from top to bottom like a traditional JPEG (we can read more about the Progressive JPEG at the following link: `https://en.wikipedia.org/wiki/JPEG#JPEG_compression`):

```
{% assign product_object = all_products["gardening-gloves"] %}
{% for image_item in product_object.images %}
  <img src="{{ image_item | img_url: "300x300", crop:
    "center", format: "pjpg" }}"/>
{% endfor %}
```

Now that we have familiarized ourselves with the parameters that are accessible using the `img_url` filter, it is time to learn more about the parameters available with the `img_tag` filter. Let's start by modifying our last example to use `img_tag` to generate the HTML tag:

```
{% assign product_object = all_products["gardening-gloves"] %}
{% for image_item in product_object.images %}
  {{ image_item | img_tag }}
{% endfor %}
```

Note that when we first mentioned `img_tag`, we used it in combination with `asset_url` to recover the URL string of the image location inside the `Assets` directory. However, since we are not accessing the `Assets` directory but the product images, whose array of URL strings we already have from using `product_object.images`, we do not need to use any additional filters besides `img_tag`.

By reviewing the results, we can see that we have successfully created the HTML `img` tag for each image within the image array. Since we haven't declared the image size, Shopify has by default resized our images to `100x100`.

`img_tag` accepts only three parameters. Contrary to `img_url`, where we can apply only the parameter we need, for `img_tag`, we need to apply all parameters in a specific order. This means that we would first have to use the other parameters if we would like to use the `class` and `alt` tag parameters.

Since we need to add all parameters in a specific order, the parameters require no representation. We only need to assign their values. The first value is alt text, inside of which we can use a fixed string value or a Liquid value, such as `image_item.alt`, to recover the actual image alt text. The second parameter we can use to assign specific classes to each image tag, while only in the third parameter can we assign the size value:

```
{% assign product_object = all_products["gardening-gloves"] %}
{% for image_item in product_object.images %}
  {{ image_item | img_tag: image_item.alt, "class1 class2",
    "300x300" }}
{% endfor %}
```

As we have had the chance to see, both the `image_url` and `img_tag` filters have their helpful parameters, and while `img_tag` is cleaner, it has limitations as we are limited in the number of attributes we can include in the generated HTML `img` tag.

Suppose we were looking to extract an image file from the `Assets` directory with the appropriate `size` parameter applied to apply it as a background image. We cannot use `img_tag`, as previously mentioned, as that would return an HTML `img` tag. We also cannot use `asset_url` alone, as `asset_url` does not accept any additional parameters, including the `size` parameter.

Similarly, as with `stylesheet_tag` and `script_tag`, we have access to the special `asset_img_url` filter, which allows us to include the `size` parameter to recover the images from the `Assets` directory:

```
{{ "ajax-loader.gif" | asset_img_url: "300x300", scale: 2,
    crop: "center" }}
```

Note that `asset_img_url` allows us to include the `size` parameter and other parameters previously available with the `img_url` filter, including `size`, `crop`, `scale`, and `format`.

So far, we have learned how to access different types of files within the `Assets` directory and generate the appropriate HTML tag for each of them. We have also learned how to work with the `img_url` filter by working through a small project of outputting the necessary elements to create a basic product gallery. While we have not gone over all of the URL and HTML filters that Liquid offers, we have now set the proper groundwork for working with HTML and URL filters, knowledge of which is essential in Liquid and will be of great help to our future work.

> **Tip:**
> For additional information on all the available HTML filters, we can refer to `https://shopify.dev/docs/themes/liquid/reference/filters/html-filters`, and for additional information on the available URL filters, we can refer to `https://shopify.dev/docs/themes/liquid/reference/filters/url-filters`.

Enhancing the product media gallery

In the previous exercise, we learned how to output the image elements needed to output the necessary elements to create a basic product gallery that contains only images. In the following project, we will learn how to use media objects and filters to create a multifunctional gallery that will support images, 3D models, and internal videos hosted on Shopify. Additionally, we will also embed external video links to some of the most popular video platforms, Vimeo and YouTube, and autogenerate the appropriate video player for both.

Most of the newest themes today already contain a product media gallery. However, many stores still use outdated theme files, so it is essential to know how to create the feature from scratch.

Let's start by navigating to the Product Gallery page we created in the previous *Building a product gallery* subsection, located under the *Working with HTML and URL filters* section, and revise the previously included code to accept additional media types besides images:

```liquid
{% assign product_object = all_products["gardening-gloves"] %}
{% for image_item in product_object.images %}
  {{ image_item | img_tag: image_item.alt, "class1 class2",
    "300x300" }}
{% endfor %}
```

Initially, we have used the product_object variable to capture the product object of the gardening gloves product, after which we have used a for tag to loop over the array of images received from product_object.images. Considering that we are dealing with various media types, we will need to use a media attribute to recover the media array and replace the image_item variable with media to keep everything cohesive:

```liquid
{% assign product_object = all_products["gardening-gloves"] %}
{% for media in product_object.media %}
  {{ media | img_tag: media.alt, "class1 class2", "300x300" }}
{% endfor %}
```

Using the `media` attribute, we have now recovered an array of all the different media objects, which can contain the following media types:

- `image`
- `external_video`
- `video`
- `model`

However, having an array of media types also means that we now have a mixed array of objects, so we need to filter them out before anything.

We will use the `case/when` tags combined with the `media_type` attribute, a part of the media object that will allow us to create a `switch` statement to recover an array of all media of a specific type. We can remind ourselves of `case/when` tags by visiting the *Controlling the flow of Liquid* section, which we can find in *Chapter 3, Diving into Liquid Core with Tags*.

Let's create a `case` tag to filter out `media_type` and write a `switch` statement for each media type:

```
{% assign product_object = all_products["gardening-gloves"] %}
{% for media in product_object.media %}
  {% case media.media_type %}
    {% when "image" %}
    {% when "external_video" %}
    {% when "video" %}
    {% when "model" %}
    {% else %}
  {% endcase %}
  {{ media | img_tag: media.alt, "class1 class2", "300x300" }}
{% endfor %}
```

With the `case/when` tags in place, we have successfully filtered out the media types and have gained access to each media type object we will need to output the media tags.

If we look at our code, we will notice that the code we have previously used to output the images in our previous example is still there. Since we now have access to the `image` object within the first `switch` statement, we can simply reposition the code inside, which will be the first step in outputting the image media files:

```
{% assign product_object = all_products["gardening-gloves"] %}
{% for media in product_object.media %}
  {% case media.media_type %}
    {% when "image" %}
      {{ media | img_tag: media.alt, "class1 class2",
          "300x300" }}
    {% when "external_video" %}
    {% when "video" %}
    {% when "model" %}
    {% else %}
  {% endcase %}
{% endfor %}
```

With `img_tag` in place, we have now successfully outputted all the image files for our product. However, we still lack tags for other media types. So, let's proceed with the external video.

The `external_video` object provides us with information about Vimeo or YouTube videos associated with a specific product. Similarly, as with image objects, to output the `external_video` media types, we will need to use `external_video_tag` to generate the necessary `iframe` element, whether for Vimeo or YouTube:

```
{% assign product_object = all_products["gardening-gloves"] %}
{% for media in product_object.media %}
  {% case media.media_type %}
    {% when "image" %}
      {{ media | img_tag: media.alt, "class1 class2",
          "300x300"}}
    {% when "external_video" %}
      {{ media | external_video_tag }}
    {% when "video" %}
    {% when "model" %}
```

```
      {% else %}
    {% endcase %}
  {% endfor %}
```

With `external_video_tag` in place, we will automatically generate an iframe with all the necessary attributes for each `external_video` media type, so let's give it a quick test:

1. To test out whether `external_video_tag` works, we will first need to include either a Vimeo or YouTube video on our product page media.

2. We will need to click on the **Products** section in our admin sidebar and navigate to the product whose media files we are currently viewing to do this. In our case, that product is gardening gloves.

3. After opening the specific product page, we will need to scroll down to the **Media** section, where in the top-right corner, we will find a drop-down link named **Add media from URL**, which we should click to reveal the drop-down options. Inside the dropdown, we will find two options, the first one allowing us to add an external image to our product, and the second one allowing us to embed either a Vimeo or YouTube video.

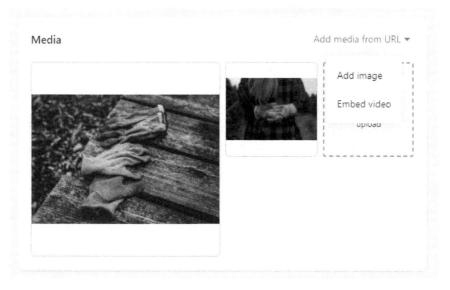

Figure 5.1 – Example of embedding external assets to a product media

4. We should proceed by clicking on **Embed video**, which will trigger a popup where we can include the URL of either a Vimeo or YouTube video. After pasting the link, click on the **Embed video** button to complete the process. After a few seconds of processing, the media video will be visible in the **Media** section.

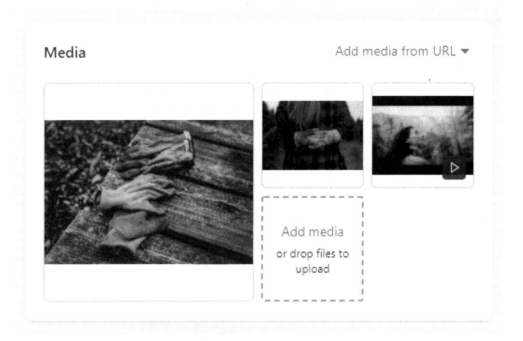

Figure 5.2 – Example of various media files on the product admin page

With the video now successfully loaded within the product media assets, all that is left is to test whether `external_video_tag` works well.

By clicking on the media video asset, we will trigger a popup with a preview on one side and the option to include alt text on the other. Additionally, in the top-right corner, we can see three icons. The first one is a trash icon, which will allow us to delete the specific media. The middle icon, represented by three dots, when clicked will reveal a drop-down menu with the **Replace thumbnail** option.

Clicking this option will allow us to upload a thumbnail visible as a poster image for the iframe video and avoid having the first frame of the video as a poster. Note that the icon represented with three dots is only visible on video assets and will not be visible on regular image assets.

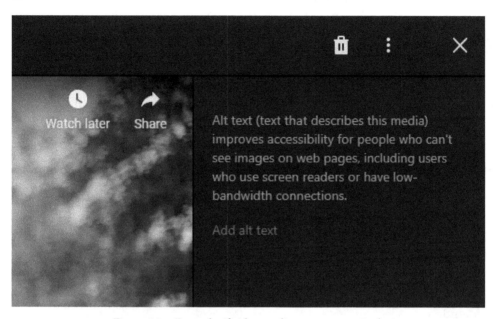

Figure 5.3 – Example of video media asset pop-up tools

Previewing the **Product Gallery** page shows us that we have correctly generated the iframe video with all the necessary attributes. First, however, let's look closely at all the attributes included with the `iframe` element:

```
<iframe frameborder="0" allow="accelerometer; autoplay;
encrypted-media; gyroscope; picture-in-picture"
allowfullscreen="allowfullscreen" src="https://www.youtube.
com/embed/neFK-pv2sKY?controls=1&amp;enablejsapi=1&amp;modest
branding=1&amp;origin=https%3A%2F%2Flearn-liquid-with-packt.
myshopify.com&amp;playsinline=1&amp;rel=0" title="Gardening
Gloves"></iframe>
```

As we can see, by default, there are many attributes included in our YouTube video embedding, such as visible controls and branding. *But what if we wanted to modify those attributes or include some new ones?*

To modify existing external video attributes or include new ones, we will need to introduce a new media filter, `external_video_url`. By including the `external_video_url` filter, in combination with `external_video_tag`, we will be able to modify any attribute of the YouTube embedding:

```liquid
{% assign product_object = all_products["gardening-gloves"] %}
{% for media in product_object.media %}
  {% case media.media_type %}
    {% when "image" %}
      {{ media | img_tag: media.alt, "class1 class2",
          "300x300" }}
    {% when "external_video" %}
      {{ media | external_video_url: controls: 0, color:
          "white" | external_video_tag }}
    {% when "video" %}
    {% when "model" %}
    {% else %}
  {% endcase %}
{% endfor %}
```

As we can see, we can include any number of attributes from the official YouTube documentation. While helpful for YouTube video embeddings, the attributes we have included will not help us when dealing with Vimeo videos, considering that Vimeo uses different attributes based on its own documentation.

In order to differentiate which embedding link belongs to YouTube and which belongs to Vimeo, we will need to introduce the `host` attribute through the `external_video` object, which returns one of two values:

- `youtube`
- `vimeo`

Using the `host` attribute in combination with an `if` statement, we can easily distinguish the two `iframe` elements and apply the appropriate attributes to each of them:

```
{% assign product_object = all_products["gardening-gloves"] %}
{% for media in product_object.media %}
  {% case media.media_type %}
    {% when "image" %}
      {{ media | img_tag: media.alt, "class1 class2",
          "300x300" }}
    {% when "external_video" %}
      {% if media.host == "youtube" %}
        {{ media | external_video_url: controls: 0, color:
            "white" | external_video_tag }}
      {% else %}
        {{ media | external_video_url: loop: 1, muted: 1 |
            external_video_tag }}
      {% endif %}
    {% when "video" %}
    {% when "model" %}
    {% else %}
  {% endcase %}
{% endfor %}
```

With the introduction of the `host` attribute, we have ensured that each type of external video embedding will receive the appropriate attributes based on its documentation.

> **Tip:**
>
> For available attributes for YouTube videos, refer to `https://developers.google.com/youtube/player_parameters#Parameters`.
>
> For available attributes for Vimeo videos, refer to `https://vimeo.zendesk.com/hc/en-us/articles/360001494447-Using-Player-Parameters`.

We have now learned how to include both types of external videos and include any attributes for each type. Now we can move on to the following `switch` statement and learn how to output a video hosted on the Shopify platform itself.

Uploading a video as a product media asset is the same as uploading an image. To upload a video, click on the **Products** section within the admin sidebar and select the product for which we are uploading the video asset. Once inside, scroll down to the **Media** section and click on the **Add Media** button to start the uploading process.

> **Important note:**
>
> Aside from external videos, uploading assets to the Shopify platform has specific limitations for all assets, including images and videos. Besides the specific types of formats that we can use, an image file cannot exceed the resolution of 20 MP, 20 MB in size, and must be in either .jpeg or .png format, whereas video files are limited to a maximum length of 60 seconds, 20 MB in size, and must be in either .mp4 or .mov video format.

Once we have uploaded the video, we can also add optional information by including alt text and a poster image in the same way as for external media.

To output the video uploaded to product media files, we will need to use video_tag, accessible through the video object:

```liquid
{% assign product_object = all_products["gardening-gloves"] %}
{% for media in product_object.media %}
  {% case media.media_type %}
    {% when "image" %}
      {{ media | img_tag: media.alt, "class1 class2",
          "300x300" }}
    {% when "external_video" %}
      {% if media.host == "youtube" %}
        {{ media | external_video_url: controls: 0, color:
            "white" | external_video_tag }}
      {% else %}
        {{ media | external_video_url: loop: 1, muted: 1 |
            external_video_tag }}
      {% endif %}
    {% when "video" %}
      {{ media | video_tag }}
    {% when "model" %}
    {% else %}
  {% endcase %}
{% endfor %}
```

By previewing the **Product Gallery** page, we will notice that in the place where the video should be, we only have a static image, a relatively small static image. If we inspect the results we have received, we will notice that we have correctly generated the HTML `video` tag. However, since we haven't enabled the controls, they have been hidden by default. Additionally, if we look closer, we will notice that the URL for the `poster` attribute within the HTML `video` tag has a small size, which is why the image on the storefront is small.

As we recall from when learning about image filters, if we try to output an image without defining the image size, Shopify will automatically resize the image to `100x100`.

With `video_tag`, we can include any number of attributes that we could use with the regular HTML `video` tag, including the `image_size` parameter, allowing us to modify the video `poster` attribute size. Let's modify `video_tag` to make the controls visible, set the poster image size, and set the video size to `300x` to match the poster size:

```
{% assign product_object = all_products["gardening-gloves"] %}
{% for media in product_object.media %}
  {% case media.media_type %}
    {% when "image" %}
      {{ media | img_tag: media.alt, "class1 class2",
            "300x300" }}
    {% when "external_video" %}
      {% if media.host == "youtube" %}
        {{ media | external_video_url: controls: 0, color:
              "white" | external_video_tag:
                    class: "youtube_video" }}
      {% else %}
        {{ media | external_video_url: loop: 1, muted: 1 |
              external_video_tag: class: "vimeo_video" }}
      {% endif %}
    {% when "video" %}
      {{ media | video_tag: controls: true, image_size:
            "300x300", width: "300x" }}
    {% when "model" %}
    {% else %}
  {% endcase %}
{% endfor %}
```

With the additional attributes included, we have now successfully generated the HTML `video` tag while retaining the ability to modify any attribute as we see fit.

The following media type that we need to look into is the 3D model, which we can output using `model_viewer_tag`, accessible through the `model` object. Simply including `model_viewer_tag` with the `media` object will automatically generate the appropriate model viewer:

```liquid
{% assign product_object = all_products["gardening-gloves"] %}
{% for media in product_object.media %}
  {% case media.media_type %}
    {% when "image" %}
      {{ media | img_tag: media.alt, "class1 class2",
            "300x300" }}
    {% when "external_video" %}
      {% if media.host == "youtube" %}
        {{ media | external_video_url: controls: 0, color:
              "white" | external_video_tag:
                  class: "youtube_video" }}
      {% else %}
        {{ media | external_video_url: loop: 1, muted: 1 |
            external_video_tag: class: "vimeo_video" }}
      {% endif %}
    {% when "video" %}
      {{ media | video_tag: controls: true, image_size:
            "300x300", width: "300x" }}
    {% when "model" %}
      {{ media | model_viewer_tag }}
    {% else %}
  {% endcase %}
{% endfor %}
```

Note that the model viewer will automatically include specific attributes by default as with the previous `media` tags. Although, if we choose so, we can easily update or even include new attributes by following the same format mentioned with the previous `media` tags.

With `model_viewer_tag` in place, we have covered all four media types and ensured that we would represent each media type with an appropriate `media` tag. However, notice that we still have one final `switch` statement without any `media` tag.

We can consider the final `switch` statement as a failsafe if, for some reason, any of the previous `switch` statements or media fail to produce results, and we will use it with `media_tag`.

The `media_tag` filter is somewhat of a special kind of filter as this particular filter will automatically generate the appropriate `media` tag for any of the four previously mentioned media types. *So, if we can use* `media_tag` *to generate all the* `media` *tags automatically, why don't we use* `media_tag` *to generate all the media files? Why did we bother to learn about all the* `media` *tags until now?*

While it is correct that `media_tag` will automatically detect each type of media type and render the appropriate `media` tag for it, by using `media_tag`, we will lose the ability to assign custom class names and attributes to specific media types. For this reason, we should never use `media_tag` as the primary option for rendering media assets. Instead, we should use it as a failsafe to correctly render the media assets on the storefront:

```
{% assign product_object = all_products["gardening-gloves"] %}
{% for media in product_object.media %}
  {% case media.media_type %}
    {% when "image" %}
      {{ media | img_tag: media.alt, "class1 class2",
          "300x300" }}
    {% when "external_video" %}
      {% if media.host == "youtube" %}
        {{ media | external_video_url: controls: 0, color:
            "white" | external_video_tag:
                class: "youtube_video" }}
      {% else %}
        {{ media | external_video_url: loop: 1, muted: 1 |
            external_video_tag: class: "vimeo_video" }}
      {% endif %}
    {% when "video" %}
      {{ media | video_tag: controls: true, image_size:
          "300x300", width: "300x" }}
    {% when "model" %}
      {{ media | model_viewer_tag }}
    {% else %}
```

```
   {{ media | media_tag }}
  {% endcase %}
 {% endfor %}
```

So far, we have learned how to output image type media files, differentiate between different types of external media assets, render the media tag for a video hosted on the Shopify platform, and generate the appropriate media tag for 3D models. Finally, with `media_tag`, we have covered every media type Shopify currently covers and have ensured that we will correctly present each media asset on our storefront.

Now that we have all the necessary assets to create the media gallery, the only thing left to do is to refine our code by including some HTML elements to format it properly. Besides the code format, we can also use some slider plugins, such as **Slick**, depending on the gallery we are looking to create. To keep everything concise and to the point, we will not be covering the media gallery's style and functionality. However, we can find the necessary suggestions on styling and functionality in the following Shopify article for those looking to test their skills by finishing the project (`https://shopify.dev/tutorials/add-theme-support-for-rich-media-3d-and-video`).

Inside the article, we can find information such as using the aspect ratio box to create responsiveness or answers to some of the frequently asked questions on functionality, such as connecting the thumbnail images to the main gallery or the variant themselves.

> **Tip:**
> For additional information on all the available media filters, we can refer to `https://shopify.dev/docs/themes/liquid/reference/filters/media-filters`.

While working on this project, we created a fundamental version of the product media gallery, allowing us to output any type of product media to any page. While it might not look impressive, the knowledge we have learned with this project has taught us how to create one of the most sought-after features today, and it will be of great help to us in creating more advanced functionalities as our knowledge grows further.

Building product accordions

In the following project, we will be learning about the string and array filters by working with and creating the product accordions feature. *But, before we proceed with the project, what exactly do string and array filters do?*

String filters are methods that allow us to manipulate the output of Liquid code or the variable itself as long as the variable is a string type, whereas array filters allow us to manipulate the output of arrays.

For this project, we will first find one product with a lengthy description. To save some time, we already included the necessary description in the **Black Armchairs** product that we previously imported from the `product-data.csv` file in *Chapter 3, Diving into Liquid Core with Tags.*

Black Armchairs

$69.99 ~~$80.00~~ [SALE]

Tax included.

ADD TO CART

BUY IT NOW

Lorem ipsum dolor sit amet, consectetur adipiscing elit. Mauris pellentesque venenatis tincidunt. Donec malesuada magna facilisis commodo convallis. Praesent at egestas lacus. Praesent dapibus mollis metus, nec iaculis tellus sodales sed. Nam commodo magna ut ornare varius. Vestibulum porta tellus et arcu iaculis, in dignissim eros convallis. Nulla facilisi.

Pellentesque quis nisl malesuada, lacinia nunc vitae, venenatis leo. Aliquam ut mauris tincidunt, mollis eros vehicula, sagittis ante. Quisque ornare tincidunt rutrum. Morbi vulputate lectus ac dui consequat viverra. Suspendisse eu egestas enim. Nunc egestas risus condimentum risus congue ullamcorper a sed urna. Nulla facilisi. Donec et ipsum tortor. Curabitur tempor mauris ac posuere tristique.

Nulla bibendum, ligula vel imperdiet luctus, ex mi dictum enim, eu laoreet nisl neque sit amet tortor. Curabitur maximus, magna et tincidunt imperdiet, libero nisi semper nibh, sed suscipit ligula erat sed massa. Nam at purus at magna ornare egestas non ac enim. Ut nibh magna, efficitur eu placerat vitae, iaculis in risus. Nulla eget iaculis nunc, vitae imperdiet elit. Morbi vel tincidunt ligula. In id mattis tellus.

Figure 5.4 – Example of a long product description

As we can see from the previous screenshot, having a lengthy product description can be quite inefficient as it takes up a lot of space. While we can easily format the product description code to include the necessary HTML tags to create product accordions, manually adjusting the code for each product would be a long process. Maintaining it is even worse. Luckily, by using string and array filters, we can easily manipulate the product description output to break it up and format it in any way we need:

1. First, let's begin by identifying the piece of code that is rendering the current product description. We can find the product description inside `product-template.liquid` under the `Sections` directory, which we are currently rendering using `product.description`:

```
<div class="product-single__description rte">
    {{ product.description }}
</div>
```

2. Since we are looking to create reusable code, the first thing we should do is to create a variable that will hold the `product.description` output, so that we can avoid calling `product.description` multiple times:

```
<div class="product-single__description rte">
    {% assign productDescription = product.description%}
</div>
```

Now that we have the variable in place, we should outline what exactly we are trying to achieve. Again, looking over the current product description, we can clearly see that we have three solid blocks, so let's say that we are looking to separate the entire product description into three or even more distinctive product accordions.

Now that we know *what*, we need to think of *how*. While we can easily hardcode the product accordion titles, such as **Description**, **Ingredients**, and **Instructions**, we are looking to create a dynamic feature that allows us to easily include any number of product accordions without modifying the code itself. We'll use the following steps to do so:

- Start by navigating to the **Black Armchairs** product on the admin page and include the words `Description`, `Ingredients`, and `Instructions` as h6 headings, one before each text block. To do this, we can use the headings to include any number of product tab titles and set the proper markings for later.

- Then, we can apply the headings using the rich text editor by simply highlighting the text we are looking to format and clicking the **A** button, which will trigger a dropdown where we can select the h6 heading we need:

Figure 5.5 – Example of applying headings inside the product description

With the headings in place, we now have proper markup that we can use to separate the text blocks into separate blocks, which we can do using the following string filter, `split`.

The split filter

The `split` filter uses a single substring as a parameter, which acts as a delimiter, dividing the string into an array whose items we can later output using array filters. *But how exactly does it work?*

```
{% assign methods = "Strings and Filters" %}
```

In this example, we have created a variable named `methods` and assigned a string message to it. Let's now use the `split` filter to divide the string message into an array and call it immediately to see the results:

```
{% assign methods = "Strings and Filters" | split: " and " %}
{{ methods }}
```

```
StringsFilters
```

As we can see, after applying the `split` filter, the substring that we used as a delimiter was removed entirely from the initial string, and we have ended up with an array result. Therefore, any value assigned to the `split` filter substring parameter will not only serve as a markup delimiter but will also automatically remove any occurrence of the substring value from the string.

After using the `split` filter, we have now modified the `methods` variable into an array. However, this is not obvious since we have also included empty space within the delimiter, so as a result, we have two words written next to each other without any space. To test whether our `methods` variable is an array, we need to run it through a loop using the `for` tag. *However, what if we wanted to avoid using a loop since we only have two items inside the array?*

This is where the `first` array filter comes to help. As we recall, we previously mentioned both the `first` and `last` filters in the previous chapter when we worked on the `Product Customization` project. Using the `first` or `last` filter will automatically recover the first or last element inside the array. Since our array only has two elements after using the `split` filter, it is a perfect fit. Otherwise, we would need to include a `for` tag to loop over the array to recover the proper values. Let's see it in action by recovering only data before the delimiter:

```
{% assign methods = "Strings and Filters" | split: " and "
    | first %}
{{ methods }}
```

```
Strings
```

We can now see that we have successfully modified the initial string type variable into an array type and have successfully recovered only the first item of the array. While this specific method might look non-important, the knowledge we have gained will be of great use in our future work.

Before we move on to our project, let's take one more example:

```
{% assign message  = "This is a short string message." |
   split: " " %}
```

In this example, we have assigned a short string inside the `message` variable. We have then applied the `split` filter and set its substring to an empty space value, meaning that we are looking to divide the initial string for every empty space occurrence. Let's create a short loop to confirm whether the `message` variable is now an array type:

```
{% assign message  = "This is a short string message." |
   split: " " %}
{% for item in message %}
   {{ item }}
{% endfor %}

This
is
a
short
string
message.
```

As we can see, by introducing the `split` filter, we have successfully divided the initial string message using the substring parameter and created an array, which we have confirmed using the `for` tag. Let's now return to our project and use our newfound knowledge to divide the product description into three separate blocks.

Since we are looking to create a dynamic feature, we will need to set the substring parameter to a value that we know will be present in each product description. Remember that we have previously added the headings, formatted using the `h6` heading. So, let's use the `split` filter and set the substring value to the opening `h6` heading:

```
<div class="product-single__description rte">
   {% assign productDescription = product.description |
      split: "<h6>" %}
</div>
```

To make sense of using the h6 heading as a substring parameter, we will first need to look over the current HTML format of our product description. If we were to inspect the product description before using the split filter, we would notice the following HTML format:

```
<h6>Description</h6>
<p>Lorem ipsum content<p>
<h6>Ingredients</h6>
<p>Lorem ipsum content<p>
<h6>Instructions</h6>
<p>Lorem ipsum content<p>
```

As the markup shows, each heading is located just above the text block, providing perfect markup. Using the opening h6 heading as a delimiter, we should have four items inside our array. Let's now see what happens when we apply the split filter using the opening h6 heading as a delimiter:

```
<div class="product-single__description rte">
  {% assign productDescription = product.description |
    split: "<h6>" %}


{% for item in productDescription %}
  <div class="product-single__description-item">
     {{ item }}
  </div>
{% endfor %}
</div>
```

Based on our markup, after applying the split filter with the opening h6 heading as a delimiter, the first item in our array should be empty, as there is no content before the first occurrence of our delimiter. The other three should contain both the heading and any content between the first and the following delimiter occurrence:

```
<div class="product-single__description-item"></div>
<div class="product-single__description-item">
<span>Description</span>
<p>Lorem ipsum content<p>
</div>
<div class="product-single__description-item">
<span>Ingredients</span>
```

```
<p>Lorem ipsum content<p>
</div>
<div class="product-single__description-item">
<span>Instructions</span>
<p>Lorem ipsum content<p>
</div>
```

Looking over the results, we can see that we were successful in extracting each product description block. However, there are still a few things to cover. For example, we will have one empty div, which we can quickly resolve by introducing the offset parameter and setting its value to 1, allowing us to skip the first iteration within the for loop. We can remind ourselves of the offset parameter by visiting the *Iterations tags* section located in *Chapter 3, Diving into Liquid Core with Tags*:

```
<div class="product-single__description rte">
   {% assign productDescription = product.description |
      split: "<h6>" %}

{% for item in productDescription offset: 1 %}
   <div class="product-single__description-item">
      {{ item }}
   </div>
{% endfor %}
</div>
```

Adding the offset parameter to the for tag will skip the first iteration, otherwise returning an empty value. Additionally, we have ensured that our array only contains the three product description blocks we initially intended. However, let's take a closer look at the current results of our array after applying the offset parameter:

```
<div class="product-single__description-item">
<span>Description</span>
<p>Lorem ipsum content<p>
</div>
<div class="product-single__description-item">
<span>Ingredients</span>
<p>Lorem ipsum content<p>
</div>
<div class="product-single__description-item">
```

```
<span>Instructions</span>
<p>Lorem ipsum content<p>
</div>
```

As we recall, by using the opening `h6` heading as a substring to the `split` filter, we will automatically remove any opening `h6` heading within the initial string. However, notice that we have not only removed the opening `h6` tag but we have also removed the closing `h6` tag.

Technically, the closing `h6` tag is still there, located after the closing `span` tag. However, since we have removed the opening `h6` tag, the browser interpreted this as an error and automatically removed the closing `h6` tag. So, instead of relying on the browser to clean up, let's use the closing `h6` tag to divide our three blocks further.

Currently, each item variable inside the `for` loop contains both the heading and the content text. By applying the `split` filter using the closing `h6` heading, we will modify the `item` variable into an array of its own, containing the heading and the content. So, instead of using another `for` loop to loop over those, let's recall the `first` and `last` filters that we previously mentioned and use them here to recover each value separately:

```
<div class="product-single__description rte">
  {% assign productDescription = product.description |
    split: "<h6>" %}


{% for item in productDescription offset: 1 %}
  <div class="product-single__description--item">
    <div class="product-single__description-title">
      {{ item | split: "</h6>" | first }}
    </div>
    <div class="product-single__description-content">
      {{ item | split: "</h6>" | last }}
    </div>
  </div>
{% endfor %}
</div>
```

By applying the `split` filter again, we have removed the leftover closing `h6` tag. Additionally, we have separated the content more clearly, allowing us to use the results to complete our mini project more easily.

At this point, we possess all the necessary elements to finalize the product accordions project. The only thing left is to include some styling and introduce the script to handle the on-input animation. To keep everything concise, we will not cover the styling and functionality within this project. However, we will provide the final expectation to help us visualize a clear goal that we should work on, as it will serve as an excellent practice for future work:

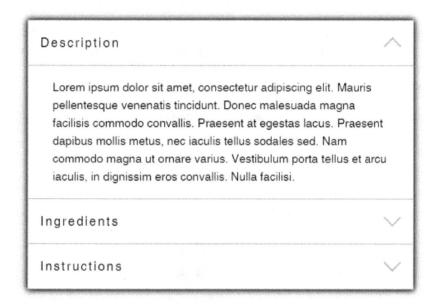

Figure 5.6 – Example of a complete product accordions project

So far, we have learned how to manipulate string variables by turning them into an array type variable and using them to create the product accordions feature. In addition, we have learned about the `split` and `first/last` filters that we included using the pipeline character. However, specific filters, such as `size`, can be used with pipeline and dot annotation, depending on the situation:

```
{% assign message  = "This is a short string message." %}

{{ message | size }}
{% if message.size >= 10 %}
  This message contains more than 10 characters.
{% endif %}
```

In the previous example, we have created a `message` variable and have assigned a short string to it. Using the `size` filter with the `message` variable, we will render the total number of characters within the message string. However, by using the size with the dot annotation, we gained the ability to use the `size` filter as part of Liquid logic:

```
31
```

```
This message contains more than 10 characters.
```

We can also use the `size` filter to improve our product accordions by including an `if` statement that will check whether the product description has more than 1 block of code and render the product accordions if it does. Otherwise, it should simply output the regular product description:

```
<div class="product-single__description rte">
  {% assign productDescription = product.description |
    split: "<h6>" %}

{% if productDescription.size > 1 %}
  {% for item in productDescription offset: 1 %}
    <div class="product-single__description--item">
      <div class="product-single__description-title">
        {{ item | split: "</h6>" | first }}
      </div>
      <div class="product-single__description-content">
        {{ item | split: "</h6>" | last }}
      </div>
    </div>
  {% endfor %}
{% else %}
  <div class="rte">{{ product.description }}</div>
{% endif %}
</div>
```

With the introduction of the `size` filter, we have made our code a lot cleaner and more optimized, as we will not run a `for` loop for a single item inside the array type variable.

As we can see, by simply using everyday filters, we can improve our code significantly. For example, *what if we wanted to split the product description and render the content into various places and not simply output them all at once?* For this, we will need to use a combination of the `split`, `first/last`, and `index` filters. We have already covered the `split` and `first/last` filters, *but what does the `index` filter do?*

The index filter

As its name suggests, the `index` filters allow us to access a specific array using its index location, starting from 0. Let's try to apply the index location to our product accordions project. Note that we will no longer have any need for the `for` tag, as we will be accessing the `productDescription` variable directly through the `index` filter:

```
<div class="product-single__description rte">
  {% assign productDescription = product.description |
    split: "<h6>" %}

  <div class="product-single__description--item">
    <div class="product-single__description-title">
      {{ item[1] | split: "</h6>" | first }}
    </div>
    <div class="product-single__description-content">
      {{ item[1] | split: "</h6>" | last }}
    </div>
  </div>
</div>
```

The previous code will return the first heading and block of text from our product description. However, we have mentioned that the index position starts at 0, *so why is position 1 returning the results of the first block?*

As we recall, since we have used the opening h6 tag as a delimiter to our `split` filter, the first item in our array is empty. Previously we skipped the first item inside the array by including the `offset` parameter, whereas now we will simply skip the *first index position*, which is 0. The downside of this method is that we will have to repeat the code to recover each block separately. However, on the plus side, we gain the flexibility of positioning them at the place of our choosing, which sometimes will be necessary.

> **Tip:**
>
> For additional information on all the available string filters, we can refer to `https://shopify.dev/docs/themes/liquid/reference/filters/string-filters`.
>
> For additional information on all the available array filters, we can refer to `https://shopify.dev/docs/themes/liquid/reference/filters/array-filters`.

We saw how somewhat insignificant filters could be a powerful tool that will allow us to create complex features that merchants regularly seek. Through this type of learning process, we gain valuable lessons by working on a real-life project, but we also learn how to deal with various types of filters, which is a lot more essential than simply listing them all.

Math and money filters

In the previous chapter, we had a chance to see money filters in action while working on the `Custom collections` project. **Money filters** are simple types of filters whose only task is to format the number value based on the currency formatting options, *but what exactly does this mean?*

To better understand, let's navigate to our admin page and click on the **Settings** button in the bottom-left corner. Consequently, click on the **General** option to open where we will be able to update the store's basic information. Once inside, scroll down until you have reached the section named **Store currency**. This is where we can change the store's default currency, which our customers will use to make their purchases. Instead of changing the store currency, let's click on the **Change formatting** button.

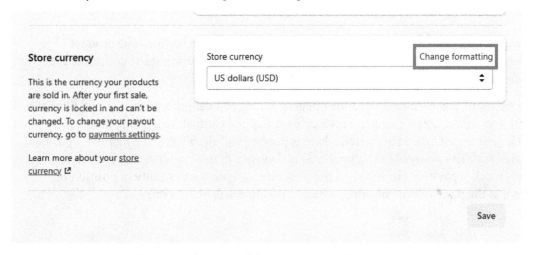

Figure 5.7 – Location of the store currency formatting

By clicking the **Change formatting** button, we will reveal additional currency formatting options where we can define the global markup for the currency formatting that the money filter will later use to format any number value.

If we were to apply the money filter to any number value, based on the formatting settings, the only change that we would see is that we would see a dollar sign before the price and USD after the price. So, let's try to modify these two fields to include some more helpful markup.

We can edit the first two fields named **HTML with currency** and **HTML without currency** by wrapping the current values inside the span tag with the money class. However, note that when writing a money class, we should not include quote marks. Otherwise, we risk breaking the currency formatting.

Store currency

This is the currency your products are sold in. After your first sale, currency is locked in and can't be changed. To change your payout currency, go to payments settings.

Learn more about your store currency 🗗

Store currency   Change formatting

US dollars (USD)   ↕

CURRENCY FORMATTING

Change how currencies are displayed on your store. {{amount}} and {{amount_no_decimals}} will be replaced with the price of your product.

HTML with currency

```
<span class=money>${{amount}} USD</span>
```

HTML without currency

```
<span class=money>${{amount}}</span>
```

Email with currency

```
${{amount}} USD
```

Email without currency

```
${{amount}}
```

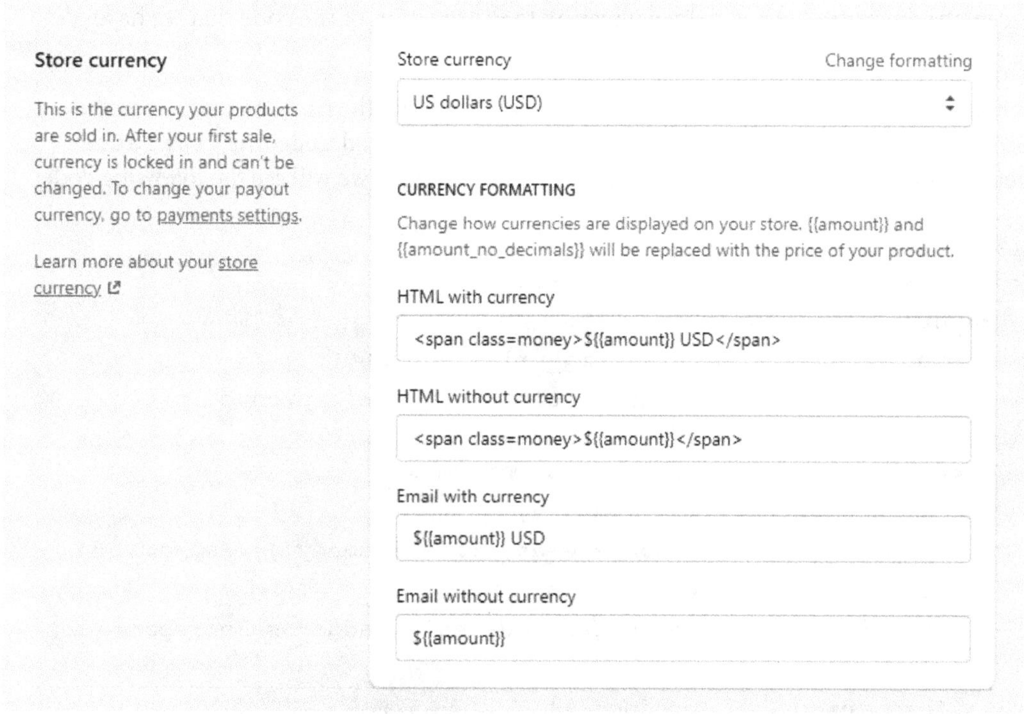

Figure 5.8 – Example of updating the store currency formatting

By including the span tag with the money class inside the currency formatting, we have now ensured that each pricing element in our store will contain the same selector, which will help if we need to update the pricing dynamically.

On the other hand, **math filters** are self-explanatory, as they allow us to perform specific mathematical tasks. Similarly, as with the string and array filters, we can chain multiple math filters inside one line, in which case, the filters will apply in order from left to right.

Now that we have familiarized ourselves with the money and math filters, let's see them in action and start working on our next mini project.

Product discount price

In the following project, we will be learning about the math and money filters by working on one of the projects we started in *Chapter 4, Diving into Liquid Core with Objects*, which is Custom collection, located under the *Global Objects* section.

We aim to create the product discount price and update the sale badge to display an actual percentage discount. Let's start by navigating to the location of the code that we have previously developed.

We created the Custom collection feature by including the collection-form. liquid snippet inside the collection.liquid file, located under the Templates directory. Inside the collection-form.liquid snippet, we will see the following code:

```liquid
{% if product.compare_at_price != blank %}
<div class="custom-collection--item">
   <a href="{{ product.url }}">
      <img src="{{ product | img_url: "300x300" }}"/>
      <p class="h4 custom-collection--title">{{ product.title
         }}</p>
      <p class="custom-collection--price">
         {{ product.price | money }}
         <span>{{ product.compare_at_price | money }}</span>
      </p>
      <span class="custom-collection--sale-badge">sale</span>
   </a>
   {% form "product", product %}
      <input type="hidden" name="id" value="{{
         product.first_available_variant.id }}" />
      <input type="submit" value="Add to Cart"/>
   {% endform %}
</div>
{% endif %}
```

With the current setup, we display both the regular price and the comparison price, so let's start by modifying the comparison price and replacing it with the actual discount price. To output the discount between the two prices, it should be relatively straightforward as we only need to subtract the regular price from the comparison price.

To subtract the regular price from the comparison price, we will need to use the `minus` math filter:

```
<p class="custom-collection--price">
  {{ product.price | money }}
  <span>{{ product.compare_at_price | minus: product.price
    }}</span>
</p>
```

Notice that with the introduction of the `minus` filter, we had to remove the `money` filter entirely. As previously mentioned, we can only use math filters if all the values are number values. If we were to include `money` filters, we would turn both values into a string due to our previously set currency formatting.

Now that we have successfully applied the `minus` filter, we need to think of a way to include the `money` filter. With the current setup, we have received a number value without any currency formatting. However, as mentioned, if we were to apply the `money` filter to any of the two values, even after `product.price`, the `money` filter would only affect `product.price`, not the final results. Consequently, the math filter would no longer work.

To resolve this, we would need to introduce a variable using an `assign` or `capture` keywords to calculate the difference between the two numbers and later call the variable with the `money` filter:

```
<p class="custom-collection--price">
  {{ product.price | money }}
  {% assign discount-price = product.compare_at_price |
    minus: product.price %}
  <span>{{ discount-price | money }}</span>
</p>
```

If we preview our collection page now, we will see that we have correctly performed both the discount price calculations and the currency formatting. So now we can move to the second part of our project, which is replacing the sale badge with an actual percentage value discount.

We can find the sale badge HTML code within the same `collection-form.liquid` snippet file:

```liquid
{% if product.compare_at_price != blank %}
<div class="custom-collection--item">
  <a href="{{ product.url }}">
    <img src="{{ product | img_url: "300x300" }}"/>
    <p class="custom-collection--price">
      {{ product.price | money }}
      {% assign discount-price = product.compare_at_price |
          minus: product.price %}
      <span>Save {{ discount-price | money }}</span>
    </p>
    <span class="custom-collection--sale-badge">sale</span>
  </a>
  {% form "product", product %}
    <input type="hidden" name="id" value="{{
        product.first_available_variant.id }}" />
    <input type="submit" value="Add to Cart"/>
  {% endform %}
</div>
{% endif %}
```

To calculate the discounted percentage, we can use the following formula, which will return the discounted percentage:

```liquid
{{ product.compare_at_price | minus: product.price | times: 100
| divided_by: product.compare_at_price }}
```

In the previous example, we had to include the calculation inside a variable so that later we could apply the money filter. However, since this time we do not need money, we can simply include the calculation with the addition of the percentage string at the end. Let's put it all together now:

```liquid
{% if product.compare_at_price != blank %}
<div class="custom-collection--item">
  <a href="{{ product.url }}">
    <img src="{{ product | img_url: "300x300" }}"/>
    <p class="h4 custom-collection--title">{{ product.title
        }}</p>
```

```
<p class="custom-collection--price">
    {{ product.price | money }}
    {% assign discount-price = product.compare_at_price |
        minus: product.price %}
    <span>Save {{ discount-price | money }}</span>
</p>
<span class="custom-collection--sale-badge">{{
    product.compare_at_price | minus: product.price |
        times: 100 | divided_by: product.compare_
                at_price }}%</span>
</a>
{% form "product", product %}
    <input type="hidden" name="id" value="{{
        product.first_available_variant.id }}" />
    <input type="submit" value="Add to Cart"/>
{% endform %}
</div>
{% endif %}
```

As we had the chance to see, similarly as with the string filters, we could chain multiple math filters easily to perform the calculations we needed. However, it is worth mentioning that besides only working with number values, math filters do not accept any type of brackets, which we would usually use to perform math calculations in specific priority.

If we need to perform calculations, we have two choices. The first one is that we let Liquid perform math calculations from left to right. On the other hand, if we need to perform calculations in a specific order, we will need to split the formula into multiple variables and combine the results later.

Tip:

For additional information on all the available math filters, we can refer to `https://shopify.dev/docs/themes/liquid/reference/filters/math-filters`.

For additional information on all the available money filters, we can refer to `https://shopify.dev/docs/themes/liquid/reference/filters/money-filters`.

While we have not covered all the math and money filter types, we have gained a solid understanding of how both math and money filters work through this project. This type of knowledge will serve as a stepping stone for the features that we will be working on as Shopify experts.

Exploring the additional filters

The **additional filters** are a set of filters that do not fit under any other filter groups. However, this does not make them any less important. While there are many types of filters that we can name here, we will only mention three of them that are the most essential as we will be using them regularly.

The default filter

As its name suggests, the **default filter** allows us to set a default value for any variable, whether it is a string, array, or hash type. Note that we can only return the default value if the variable returns `nil`, `false`, or an empty string. If the variable contains whitespace characters, we will not be able to return the `default` value:

```
Hello {{ customer.name | default: "customer" }}
```

By introducing the `default` value in the previous example, we have ensured that we will not end up with a broken string even if the customer has not provided us with their name. Additionally, we also make our code look a lot cleaner. Without the `default` filter, we would have to use an `if` statement to check whether `customer.name` exists, and depending on the results, output the value.

The t (translation) filter

The **t filter** is a translation key that allows us to access the currently active store language file under the `Locales` directory. If we navigate to the `Locales` directory, we will notice a large number of files. However, one of them will contain the string default, which is the currently active language in our store.

Let's look over one of our previous projects, `Custom collections`, which we worked on in *Chapter 4*, *Diving into Liquid Core with Objects*, in the *Working with global objects* section, where we developed a submit button whose value we hardcoded with the text **Add to Cart**. This works well for now, *but what if we changed our store language?* We would have to update any occurrence of the **Add to Cart** string manually for the entire theme.

Using the t filter, we can update any string value through our entire theme by updating a single value. The first thing that we need to do is to define the three-level JSON inside the currently active language file using the Shopify naming and grouping guidelines. Since we are looking to modify the submit button of a product, we can set the first level to product, the second to something more specific, which is the form itself, and finally, the third level points to the string we are looking to translate:

```
{
  "product": {
    "form": {
      "submit": "Add to Cart"
    }
  }
}
```

Once we have successfully created the JSON inside the language file, all that is left is to use the t filter to read the translated key value and render it on the storefront:

```
{{ product.form.submit | t }}
```

By implementing a t filter on all product forms, we will gain the ability to automatically translate all the strings without having to search for them throughout our files manually. Additionally, using the same JSON naming and grouping inside the other language files will allow us to quickly translate our entire theme by simply changing the store language.

The t filter is a powerful tool that allows us to pass multiple arguments by separating them with commas and interpolating them:

```
{{ header.general.customer | t: customer: customer.name }}
```

In the previous example, we are trying to access the customer string inside the language file. However, we have also passed the customer.name value as an argument, which we have then interpolated inside the language JSON file:

```
{
  "header": {
    "general": {
      "customer": "Welcome {{ customer }}!"
    }
  }
}
```

Besides providing us with the ability to interpolate the variables, we can also escape the translated content, include the HTML in translation keys, and pluralize translation keys, making it a pretty powerful tool at our disposal.

> **Tip:**
>
> For additional information on the t filter naming and grouping guidelines, we can refer to `https://shopify.dev/tutorials/develop-theme-localization-organize-translation-keys`.
>
> For additional information on all the available arguments for the t filter, we can refer to `https://shopify.dev/tutorials/develop-theme-localization-use-translation-keys`.

Once we have created the necessary translation keys, we can also update the translation by navigating to our admin page and clicking on **Themes** located inside the expanded **Online store** option within the sidebar. Clicking on the **Actions** button on the theme file will trigger a dropdown, whereby clicking on **Edit Language** will quickly update any translation inside the language JSON file.

The JSON filter

The JSON filter, as its name suggests, allows us to convert strings into JSON, and more importantly, will make Liquid code readable by JavaScript:

```
var product_JSON = {{ product | json }};
var cart_JSON = {{ cart | json }};
```

Note that when using the JSON filter on Liquid output, there is no need to include quotations marks, as the JSON filter will include them automatically. However, note that specific values, such as the `inventory_quantity` and `inventory_policy` fields, are not something that we can return via JSON, as Shopify has deprecated these fields due to security reasons.

> **Tip:**
>
> For additional information on all the available additional filters, we can refer to `https://shopify.dev/docs/themes/liquid/reference/filters/additional-filters`.

Summary

In this chapter, we have learned how using something trivial such as a filter to manipulate different data types can create powerful features. We have learned how using URL and HTML filters can provide us with access to the various types of assets throughout Shopify and help us generate them in the storefront using their respective HTML tags.

Working on the product media gallery project has provided us with a deeper understanding of media objects and filters, which every developer needs to be familiar with. The product accordions project taught us how to easily manipulate data using string and array filters to create unique page content elements that are clean and easily maintainable. Moving on to the math and money filters, we have gained much-needed insight into performing complex calculations through Shopify and formatting the prices according to the currency formatting set in our store.

Lastly, we learned about the additional filters, which provided us with essential knowledge on how we can assign default value variables with no assigned values and make Liquid code readable by JavaScript. By understanding how to use the translation keys, we now have the necessary knowledge to quickly update any value across our theme without manually updating each value.

The knowledge that we have attained through this chapter will be especially helpful in the next chapter, where we will learn more about the JSON settings and how we can use them to create settings that merchants can access using the theme editor.

Questions

1. Suppose that we have an array named `product_handles` with handles of 30 products. What issue in the following code would prevent us from outputting the images of all 30 products successfully?

```liquid
{% for handle in product_handles %}
  {% assign product_object = all_products[handle] %}
  {% for image_item in product_object.images %}
    <img src="{{ image_item | img_url }}"/>
  {% endfor %}
{% endfor %}
```

2. Why is only using the `model_viewer_tag` tag not recommended when creating the product media gallery?

```
{% for media in product_object.media %}
   {% case media.media_type %}
      {{ media | model_viewer_tag }}
   {% endcase %}
{% endfor %}
```

3. Which filter could we use if we were looking to access an item at a specific location inside the array?

4. What filter can we use to quickly update any occurrence of a string value inside the theme files?

Practice makes perfect

Project 3

In one of our previous exercises, we learned how to create basic and complex product galleries by outputting all types of product media types. *However, what if we only needed to output several images related to the specific variant at a time?*

For our third project, we will be working on rendering a product media gallery with distinctive markings that will allow us to show only the thumbnails of the currently selected variant.

Here are the instructions for the assets:

1. Create a new page template using the name `variant-thumbnails.liquid`.

2. Create a new page named `Product Variant Thumbnails`.

3. Create a new layout file with the name `alternate.liquid`.

4. Create a new product with at least three color variants and upload at least three media assets to represent each variant. When creating a new product, we should set the product status to Active and not leave it to Draft. Otherwise, we will not be able to access it later.

5. Create two new asset files, a stylesheet with the name `custom.css` and a script file with the name `script.js`.

6. Create a new snippet file with the name `custom-media`.

The following are the steps for the assignment:

1. Edit the new layout file by including two asset files that we have previously used throughout this project.

2. Include the layout file to the previously created page template and assign the new page template to the newly created page.

3. Navigate to the **Products** section inside the admin page, find the product that we have previously created, and edit the alt text of each image to include the name of the variant in lowercase and the actual image alt text, all separated with an underscore. For example, `red_This` is an image description. Repeat the process for all images. Note that each variant should have at least three images to it.

4. Inside `variant-thumbnails.liquid`, create a variable that we will use to access the previously created product object using the product handle.

5. Using a `for` loop, access the product media object, and pass the media object inside the `custom-media` snippet as a parameter.

6. Inside the snippet, using the case-control flow tag and the object that we have passed as a parameter, create code that will output each media file type.

7. Edit each `media` tag to include three attributes. The first one, `data-variant`, will contain the first part of the media alt text before the underscore. The second attribute, `alt`, will contain the second part of the media alt text. Finally, the third attribute, named `index`, will contain the index position value of the media file. If we are having trouble calling the `forloop` object's index value directly inside the snippet, we might want to pass it as a parameter through the snippet.

With the attributes in place, we should have all the necessary elements needed to filter out product media files and only display the thumbnails related to the currently selected color variant using the `data-variant` attribute value.

For those looking to finalize the project, follow these steps:

1. Apply the necessary styling inside `custom.css` and the necessary script inside `script.js` that will hide or display only the media files with the correct `data-attribute` value.

2. Additionally, we can create a new `for` loop that will serve as a primary media gallery.

3. Include any type of slider plugin, such as a Slick slider, and connect it to the primary media gallery.

4. Clicking on the thumbnail media asset should automatically make the primary slider scroll to the selected asset.

Section 3: Behind the Scenes

This section teaches us how to create different settings using JSON, allowing merchants to customize their themes easily through their theme editor. We will also learn about the newly introduced JSON type templates, which, combined with metafields, will allow us to create unique and easily configurable content for any number of pages. Lastly, we will dive into the Shopify Ajax API and learn how we can implement advanced functionalities and make a store more dynamic.

This section comprises the following chapters:

- *Chapter 6, Configuring the Theme Settings*
- *Chapter 7, Working with Static and Dynamic Sections*
- *Chapter 8, Exploring the Shopify Ajax API*

6
Configuring the Theme Settings

In previous chapters, we have been learning about Shopify as a platform, familiarizing ourselves with Liquid basics, and using the Liquid core to create various features on the storefront. However, unless the store owners are developers, they will not have much control over any features we create for them.

In this chapter, we will be learning how we can use JSON to create settings that are accessible through the theme editor, which will allow the store owners to easily customize the theme without making the code adjustments throughout the theme. We will cover the following topics in this chapter:

- Exploring JSON settings
- Learning about the input setting attributes
- Basic and specialized input types
- Organizing the theme editor
- Glancing at the deprecated settings

By the time we complete this chapter, we will have gained a deeper understanding of the importance of JSON, and how we can use it to create theme settings that are accessible across the entire theme on any page. We can use these settings to modify the CSS values, change the content on certain features, and even use the settings to enable or disable a particular feature altogether. By learning how to use JSON to create these settings, we will take another step toward creating a genuinely dynamic and customizable feature storefront, which is what Shopify is.

Technical requirements

While we will explain each topic and have it presented with the accompanying graphics, we will need an internet connection to follow the steps outlined in this chapter, considering that Shopify is a hosted service.

The code for this chapter is available on GitHub at `https://github.com/ PacktPublishing/Shopify-Theme-Customization-with-Liquid/tree/ main/Chapter06`.

The Code in Action video for the chapter can be found here: `https://bit.ly/3nLQgMf`

Exploring JSON settings

In *Chapter 1*, *Getting Started with Shopify*, we briefly mentioned the `Config` directory, where we can define and manage the global JSON values across the entire theme. Let's remind ourselves of the two essential `.json` files that we can find within this directory:

- The `settings_schema.json` file allows us to create and manage the content inside the theme editor on our theme, which we can reference throughout the entire theme file.

- The `settings_data.json` file, on the other hand, records all the options defined in our schema file and saves their values. We can consider this file as our theme database, which we can manage by updating the theme settings through the theme editor or by directly editing the values inside the `settings_data.json` file.

We can group the global settings options into different categories for more straightforward navigation, which we can do using the `name` and `settings` attributes:

```
{
  "name": "Category",
  "settings": [

  ]
}
```

As we can see, as with the `.json` files in the `Locales` directory mentioned in the previous chapter, the `settings_schema.json` file has a specific format that we must honor. Using the `name` attribute, we can set the name of the category and the `settings` attribute will contain the array of settings that the category will contain:

```json
{
    "name": "Category",
    "settings": [
        {
            "type": "color",
            "id": "store_background_color",
            "label": "Background Color",
            "default": "#ffffff"
        },
        {
            "type": "color",
            "id": "store_border_color",
            "label": "Border Color",
            "default": "#cccccc"
        }
    ]
}
```

In the previous example, we have included two types of color settings, one that will control the background color of our store and one that will be in charge of setting the border color we will use across our store. As we can see, we have enclosed each `settings` option within the curly brackets and separated it by commas. However, note that the last `settings` option inside the category does not have a comma.

> **Important note:**
> Including a comma after the last attribute within the `settings` block, or including the comma after the last `settings` block in the category, will result in an error and we will not be able to save our work.

Now that we have defined the global settings, we need to learn how to access them and recover their values. We can recover the value of any global input setting using the `settings` keyword and the input's ID, whose value we are looking to recover, separated by a dot and encapsulated by double curly braces:

```
{{ settings.store_background_color }}
{{ settings.store_border_color }}
```

We have now had the chance to see how we can define input setting options and read their values, but where exactly are we rendering this option and how can we modify it?

We can access the theme editor by navigating to the **Online theme** section in the admin section and clicking on the **Customize** button on the theme we are looking to customize:

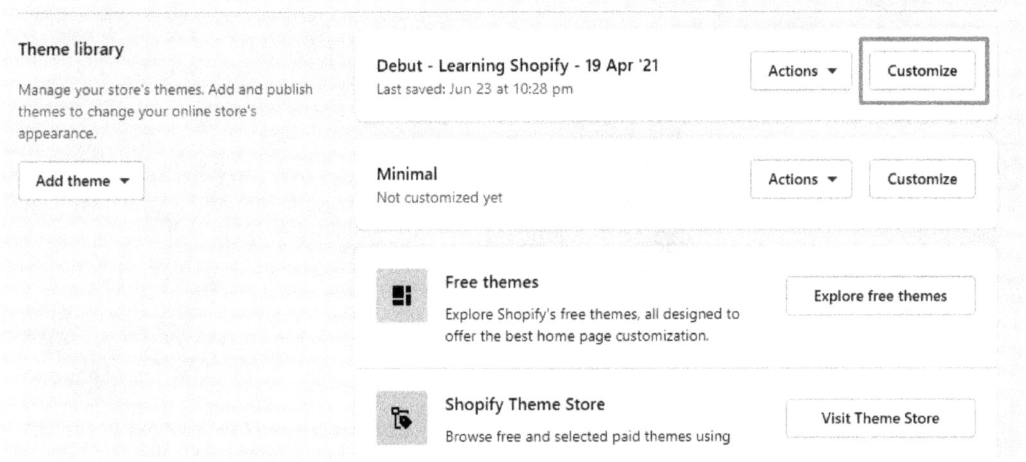

Figure 6.1 – Example of accessing the theme editor through the Online theme section

Note that changes made throughout the theme editor are theme-specific, so we should remember to click the **Customize** button on the theme we are looking to customize.

Alternatively, we can access the theme editor through the code editor by clicking on the **Customize theme** button, located in the top-right corner:

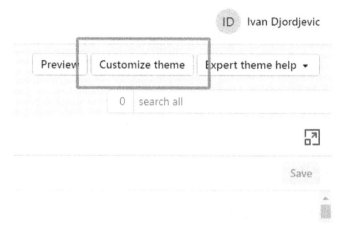

Figure 6.2 – Alternative way of accessing the theme editor through the code editor

Once inside the theme editor, we will see a list of category options within our sidebar. However, most of these options are a part of the sections and blocks that we will be learning more about in the following chapter. For now, we can access the global settings that we define throughout the settings_schema.json file by clicking on the **Theme settings** button located in the bottom-right corner:

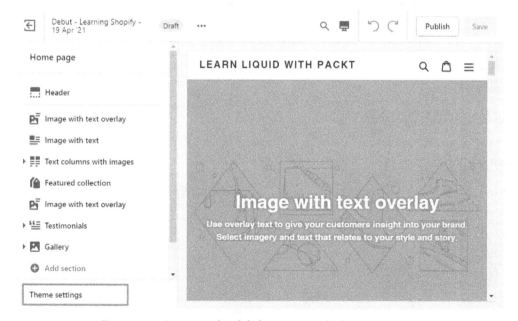

Figure 6.3 – Accessing the global settings inside the theme editor

Inside **Theme settings**, we will find all the categories and input settings that we have defined in the `settings_schema.json` file, where each input set contains a set of attributes. While some of them are required, others are optional. Let's learn more about them.

Learning about the input setting attributes

Each input settings option can contain the following five attributes, which are also called standard attributes:

- As its name indicates, the `type` attribute allows us to set the type of the input setting, which can be either a basic type or a specialized type. The `type` attribute is mandatory.

- The `id` attribute is another mandatory attribute that we will use later to access and read the setting value.

- The `label` attribute allows us to set the label of the input settings inside the theme editor. The `label` attribute is mandatory.

- The `default` value serves as a failsafe and allows us to set a default option for the input settings. However, it is not mandatory.

- The final attribute, `info`, allows us to include an additional clarification regarding the input settings and is also not mandatory.

While most of the input settings will contain only the previously mentioned attributes, depending on the input type, there will be cases where we will need to include some additional attributes.

We have previously mentioned that the type attributes allow us to choose between two different types of input settings, basic and specialized, but what exactly are they?

Basic input types

The basic input type is a set of options that allow us to include various types of input settings within our theme editor. Under the basic category, we can use the following options:

- `checkbox`
- `number`
- `radio`
- `range`
- `select`
- `text`
- `textarea`

As we have mentioned previously, most of the input settings will contain only the standard attributes. However, some of the specialized inputs and even basic inputs will require additional attributes. Let's now look into each input type, learn how to use it, and what type of results we can expect.

The checkbox input

The checkbox type of input, as its name suggests, is a Boolean type of field that allows us to create a checkbox option within the theme editor:

```
{
  "type": "checkbox",
  "id": "enable_popup",
  "label": "Enable popup",
  "default": true
}
```

As we can see, the checkbox input contains three mandatory and one optional attribute whose value we have set to true. Otherwise, if we were to remove the default attribute, the default state of the checkbox will be false. The following screenshot shows us an example of the checkbox input type:

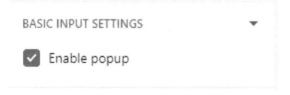

Figure 6.4 – Example of the checkbox basic input type

We can use the Boolean input type to toggle features on and off, which we can achieve by taking the checkbox input value and checking its current state using an if statement:

```
{% if settings.enable_popup == true %}
{% endif %}
```

By checking whether the checkbox input value is equal to true, we have created a simple feature that will enable or disable a certain feature from the store easily, as the code block inside the statement will only render if the statement is true:

```
{% if settings.enable_popup != blank %}
{% endif %}
```

Note that we can achieve the same results by comparing the checkbox input value against the `blank` variable.

The number input

The `number` type of input is the newest addition from Shopify and, as its name suggests, it is a number type of field, which allows us to create a number selector input inside the theme editor. In addition to the standard attributes, we can also use an optional `placeholder` attribute, which allows us to include a placeholder value for the text input:

```
{
    "type": "number",
    "id": "number_of_products",
    "label": "Number of products",
    "placeholder": "24"
}
```

Note that the `number` type of input can only contain a number value. In the following screenshot, we can see an example of the `number` input type:

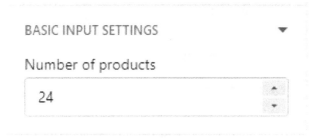

Figure 6.5 – Example of the number basic input type

Once we have the number input defined, we can access it by pairing the `settings` keyword and the ID of the text input:

```
{{ settings.number_of_products }}
```

The `number` input value will always return a number value, unless it is empty, in which case it will return an `EmptyDrop` value. We can remind ourselves of `EmptyDrop` by revisiting *Chapter 2*, *The Basic Flow of Liquid*, and checking the *EmptyDrop* subsection inside the *Understand the types of data* section.

The radio input

Using the `radio` input type, we can output a `radio` option field, which allows us to have a multi-option selection. The `radio` input uses standard attributes with the addition of the `options` attribute, which is mandatory.

The `options` attribute accepts an array of `value` and `label` attributes, which are mandatory:

```
{
  "type": "radio",
  "id": "heading_alignment",
  "label": "Heading alignment",
  "options": [
    {
      "value": "left",
      "label": "Left"
    },
    {
      "value": "center",
      "label": "Center"
    },
    {
      "value": "right",
      "label": "Right"
    }
  ],
  "default": "center"
}
```

Note that we need to set the `default` attribute to one of the values previously defined inside the `options` array. Otherwise, if the `default` attribute is not defined, the first radio will be selected by default. Here is an example of the `radio` input type:

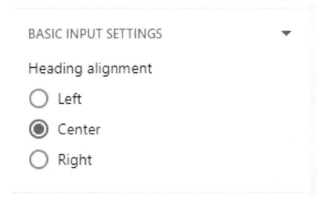

Figure 6.6 – Example of the radio basic input type

Once we have the `radio` input defined, we can access it by pairing the `settings` keyword and the ID of the radio input:

```
.heading {
    text-align: {{ settings.heading_alignment }};
}
```

The radio button value will always return a string value.

The range input

With the `range` input type, we can create a range slider field. Compared to the previous inputs, the `range` input has four additional attributes and one change to the standard attributes. We can list the additional attributes in the following way:

- The `min` attribute allows us to set the minimum value of the range input. The `min` value is mandatory.

- The `max` value is also a mandatory attribute that allows us to set the maximum value of the range input.

- The `step` value allows us to set the increment value between the steps of the slider. The `step` slider is mandatory.

- The fourth and final additional attribute, unit, is an optional attribute, which allows us to set the visual unit, such as px, for the range slider value. Note that the unit attribute only accepts up to three characters and will output the px purely visually inside the theme editor. The actual value will return a number value without the unit.

One additional change to the standard set of attributes is that the default attribute is now mandatory:

```
{
    "type": "range",
    "id": "logo_size",
    "min": 120,
    "max": 220,
    "step": 1,
    "unit": "px",
    "label": "Logo Size",
    "default": 140
}
```

Note that the min, max, step, and default attributes are a number type of value. Including a string value in any of these attributes will result in an error.

When compared to the previous attributes, the range attribute comes with a few rules that we must follow. The first rule is that the default value must be a value between the min and max values. The second and more important rule is that each range slider can have a maximum of 100 steps, but what exactly does this mean?

For example, in the previous example, we have set a min value of 120 and a max value of 220. Since we have set the step value to 1, we have precisely 100 steps between the two values.

On the other hand, if we set the max value to 320, we would also have to update the unit value to 2 to retain the 100 steps between the min and max value. In the following screenshot, we can see the example of a range input type:

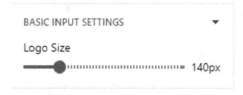

Figure 6.7 – Example of the range basic input type

Once we have the `range` input defined, we can access it by pairing the `settings` keyword and the ID of the `range` input. However, remember that the `unit` attribute is purely visual. Since the `range` input returns a number value, we would need to include the `unit` value within the stylesheet manually:

```
.logo-img {
    max-width: {{ settings.logo_size }}px;
}
```

This is one of the reasons why we should always include the `default` attribute when defining styling settings, or at least to wrap the entire CSS line within a statement that checks whether the value exists. Otherwise, we risk breaking the stylesheet if the value of the input is not defined.

The select input

The `select` input type allows us to create a drop-down selector field. In addition to the standard set of attributes, the `select` input type has two additional attributes. We can list the additional attributes in the following way:

- The `options` attribute, similar to the `range` input, allows us to create an array of `value` and `label` attributes to define the drop-down options. Both options and `value`/`label` inside the array are mandatory attributes.

- The `group` attribute is an optional attribute that allows us to group different options inside the dropdown.

Like the `radio` input type, if we do not define the `default` attribute, the first option is selected by default:

```
{
    "type": "select",
    "id": "font_family",
    "label": "Font Family",
    "options": [
        {
            "value": "raleway-light",
            "label": "Raleway - Light",
            "group": "Raleway"
        },
        {
```

```
    "value": "raleway-regular",
    "label": "Raleway - Regular",
    "group": "Raleway"
  },
  {
    "value": "playfair-display-regular",
    "label": "Playfair Display - Regular",
    "group": "Playfair Display"
  }
 ],
 "default": "playfair-display-regular"
}
```

Since the `select` input returns a string value, one of the most common uses is to include a custom font family within the theme, as we had the chance to see in the previous example:

Figure 6.8 – Example of the select basic input type

Additionally, by using the `group` attribute, we have successfully grouped all options that belong to the same family:

```
.heading {
  font-family: {{ settings.font_family }};
}
```

Shopify also has an extensive font library that we can use and a specialized input that provides us with access to the mentioned library, which we will learn shortly. However, suppose we are looking to include a custom font that is not available within the Shopify font library. In that case, we will need to include it using a custom solution by including it through a `select` type dropdown.

The text input

The text type of input, as its name suggests, is a string type of field that allows us to create a single-line text option within the theme editor. In addition to the standard attributes, we can also use a placeholder attribute, which allows us to include a placeholder value for the text input:

```
{
    "type": "text",
    "id": "header_announcement",
    "label": "Header Announcement",
    "placeholder": "Enter a short announcement.",
    "default": "Spend more than 100$ to qualify for a 10%
        discount!"
}
```

Note that the text type of input can only contain the string value and cannot include any HTML tags. In the following screenshot, we can see an example of the text input type:

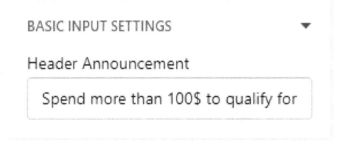

Figure 6.9 – Example of the text basic input type

Once we have the text input defined, we can access it by pairing the settings keyword and the ID of the text input:

```
{{ settings.header_announcement }}
```

The text input value will always return a string value, unless it is empty, in which case it will return an EmptyDrop value.

The textarea input

The `textarea` type of input works similarly to the text input, with the only difference being that `textarea` is a multi-line text field as compared to the `text` input, which is a single-line field. In addition to the standard attributes, we can also use a `placeholder` attribute to include a placeholder value for the `textarea` input:

```
{
    "type": "textarea",
    "id": "header_announcement_textarea",
    "label": "Header Announcement",
    "placeholder": "Enter a short announcement.",
    "default": "Spend more than 100$ to qualify for a 10%
        discount!"
}
```

Similarly, as the `text` input value, the `textarea` input value will always return a string value, unless it is empty, in which case it will return an `EmptyDrop` value. We can see the example of the `textarea` input in the following screenshot:

Figure 6.10 – Example of the textarea basic input type

With `textarea`, we have now covered all of the basic input settings, which has set us on a solid path to understanding and working with theme editor `.json` files. However, to truly be able to say that we have a working knowledge of `.json` files, we will also need to learn about the specialized input settings as well.

Specialized input settings

The specialized input type is a set of specialized options that don't allow us to include various types of input settings within our theme editor but also provide us with access to various Liquid objects easily. Under the specialized input, we can use the following options:

- `richtext`
- `html`
- `linklist`
- `liquid`
- `color`
- `url`
- `video_url`
- `image_picker`
- `font_picker`
- `article`
- `blog`
- `collection`
- `page`
- `product`

Similar to the basic input types, most of the input settings will only contain the standard attributes. However, some of the inputs will require additional attributes. By learning about all the different types of specialized inputs, we will learn how to use them and what type of results we can expect, which will help create complex sections that we will learn about in the next chapter.

The richtext input

The `richtext` input type is similar to the basic `textarea` type of input as they both output a multi-line text field. The major difference is that `richtext` also provides us with some basic formatting options:

- Bold
- Italic
- Underline
- Link
- Paragraph

The second difference is that while `text` and `textarea` input values return a clean string, `richtext` will always return a string value formatted as a paragraph, encapsulated inside the HTML `<p></p>` tag. The following screenshot shows us an example of the `richtext` input:

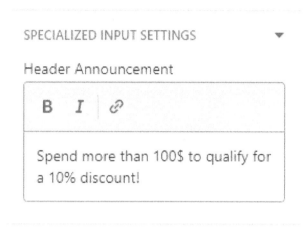

Figure 6.11 – Example of the richtext specialized input type

Additionally, using the formatting options that `richtext` provides will automatically update the string value with the respective HTML tags.

> **Important note:**
> Using the formatting options will automatically apply the necessary HTML tags to the `richtext` string value. However, we cannot manually include any HTML tags within the theme editor's `richtext` field.

While the `default` attribute is not mandatory, if we decide to use it, we will have to include the `<p></p>` tags inside the `default` attribute value. Otherwise, we will receive an error:

```
{
    "type": "richtext",
    "id": "header_announcement_richtext",
    "label": "Header Announcement",
    "default": "<p>Spend more than 100$ to qualify for a 10%
        discount!</p>"
}
```

Note that `richtext` does not accept any `placeholder` attributes compared to `textarea`, which accepts it.

The html input

The `html` input type is a multi-line text field that, as its name suggests, allows us to include HTML markup within the input field. In addition to the standard attributes, the `html` input type also accepts an optional `placeholder` attribute:

```
{
    "type": "html",
    "id": "google_analytics",
    "label": "Google Analytics",
    "placeholder": "Paste the Google Analytics code here."
}
```

While the `html` input will accept most of the HTML tags, Shopify will automatically remove the following three tags:

- `<html>`
- `<head>`
- `<body>`

The `html` type input value will always return a string value or an `EmptyDrop` value if it is empty. We can see the example of the `html` input inside the following screenshot:

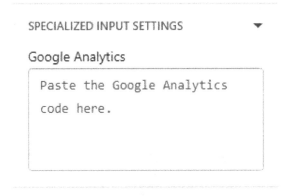

Figure 6.12 – Example of the html specialized input type

Note that while we can include HTML code inside the `html` type input, we cannot include Liquid code inside the field as it will get processed as a simple string.

The linklist input

Using the `link_list` input type, we can create a special menu picker type of field that allows us to output one of the store navigation menus. Note that we can only see the menus we have previously created inside the admin **Navigation** section, located under the **Online store** section:

```
{
    "type": "link_list",
    "id": "header-menu",
    "label": "Header Menu",
    "default": "main-menu"
}
```

Note that while the `default` attribute is optional, it only accepts two specific values, `main-menu` and `footer`. Let's see how the `linklist` input looks in the following screenshot:

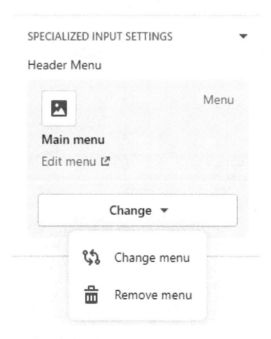

Figure 6.13 – Example of the linklist specialized input type

Using the settings and the ID to retrieve the `link_list` value will return a `linklist` object, which we can use to build the navigation menu. If we recall, in *Chapter 4, Diving into Liquid Core with Objects*, we have used the following code to output a navigation menu with a specific `indoor-navigation` handle:

```
{% assign collection-menu = linklists.indoor-navigation %}
    {% for link in collection-menu.links %}
{% endfor %}
```

And if we wanted to use different navigation, we would have to manually update the menu navigation handle. However, considering that we can now recover the `linklist` object directly from the theme editor, we can update the previously hardcoded `linklist` object value and replace it with a dynamic one:

```
{% assign collection-menu = settings.header-menu %}
    {% for link in collection-menu.links %}
{% endfor %}
```

Note that if the `link_list` type input does not have a `default` attribute or we have not yet selected the menu, we will receive a `blank` value as the return value.

The liquid input

The `liquid` input type is also a recent addition from Shopify, and it allows us to include both HTML markup and limited Liquid code, which makes it a pretty powerful tool. Here is an example code for this input:

```
{
    "type": "liquid",
    "id": "liquid_block",
    "label": "Liquid block"
}
```

Note that any unclosed HTML tag will automatically close when we save the settings. In the following screenshot, we can see an example of the `liquid` input type:

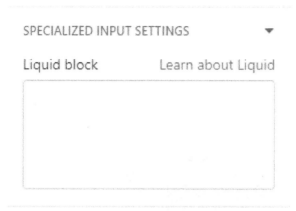

Figure 6.14 – Example of the liquid specialized input type

The `liquid` input type provides us with access to all global objects, page-based objects, and standard tags and filters.

Note that the `liquid` type input value will always return a string value or an `EmptyDrop` value if empty.

The color input

As its name suggests, the `color` input type allows us to create a color picker type input to update the store's color palette easily:

```
{
    "type": "color",
    "id": "store_background_color",
    "label": "Background Color",
    "default": "#ffffff"
}
```

We can access the values of the `color` input type by combining the `settings` keyword and the ID of the input:

```
body {
    background-color: {{ settings.store_background_color }};
}
```

Besides being able to enter the hex color manually inside the `color` field, we can trigger an actual color palette where we can select the desired color tone:

Figure 6.15 – Example of the color specialized input type

Note that while the `default` attribute is optional, we should always either include the default value or wrap the entire CSS line within the statement to check whether the input value isn't empty. If we neglect to include either of the two, in the event that the `color` input value is not defined, we may end up with a broken stylesheet.

The url input

The `url` type input provides us with a special URL entry field where we can either manually enter the external URL or use a series of dropdowns to select the path to one of the following resources:

- Articles
- Blogs
- Collections
- Pages
- Products
- Policies

We can also include a link to sites outside of our store by simply pasting the URL and consequently clicking on the link inside the drop-down menu confirming the selection:

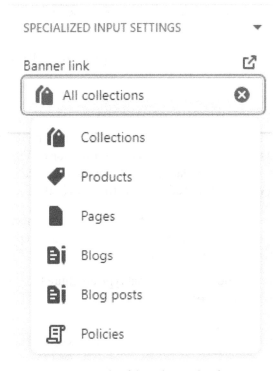

Figure 6.16 – Example of the url specialized input type

Note that while the `default` attribute is optional, it only accepts two specific values – /
collections and /collections/all:

```
{
    "type": "url",
    "id": "banner_link",
    "label": "Banner link",
    "default": "/collections"
}
```

We can access the value of the `url` input type by combining the `settings` keyword and
the input's ID, which we can then include as the `href` parameter to the HTML `<a>` tag:

```
<a href="{{ settings.banner_link }}"></a>
```

Considering that the HTML `<a>` tag is hardcoded, we cannot make any dynamic
modifications to the tag using the `url` type input. However, we can introduce an
additional input type, such as `checkbox`, which we can use to display or hide the
`target="_blank"` attribute.

Note that the `color` type input value will always return either a string value or nil if the
value is not defined.

The video_url input

The `video_url` type input provides us with a special URL entry field where we can
manually enter the external URL of a video from either YouTube or Vimeo and extract their
IDs for later use. In addition to the standard set of attributes, the `video_url` input type has
two additional attributes. We can list the additional attributes in the following way:

- The `accept` attribute is a mandatory array type attribute where we can define
  the different types of providers whose video URL we will accept. Valid values are
  `youtube`, `vimeo`, or a combination of both.

- The `placeholder` attribute is an optional type attribute that allows us to include a
  placeholder value for the `video_url` input.

The following code shows the `video_url` type input with the preceding attributes:

```
{
    "type": "video_url",
    "id": "banner_video_url",
    "label": "Video url",
    "accept":  [
      "youtube",
      "vimeo"
    ],
    "placeholder": "Enter the YouTube or Vimeo video URL."
}
```

After including the video URL to one of the two video platforms, we will see the video name and first frame, allowing us to confirm that we have the correct video URL:

Figure 6.17 – Example of the video_url specialized input type

We can access the value of the `video_url` input type by combining the `settings` keyword and the input's ID, which will return the URL that we have previously included in the theme editor:

```
{{ settings.banner_video_url }}
```

Compared to the previous input types, which return a single value, `video_url` allows us to access two additional parts of the video URL:

- The `id` attribute, which allows us to recover only the video ID

- The `type` attribute, which allows us to easily identify whether the video is from YouTube or Vimeo platforms:

```
{{ settings.banner_video_url.id }}
{{ settings.banner_video_url.type }}
```

Note that whether we are trying to return a complete URL or only its `type`/`id`, the returned value will always be a string type, unless the `video_url` value is not defined, in which case the returned value will be nil:

```
_9VUPq3SxOc
```

```
youtube
```

Now that we have recovered the video information, all that is left is to include are the returned values inside the `iframe` embed for each respective platform. Note that we will need to include a separate `iframe` embed for each platform, which should not be a problem as we can easily identify which platform the video belongs to using the `type` attribute.

Note that since `video_url` returns a string type URL, it is impossible to generate the necessary `iframe` embed using the media object we mentioned in one of the previous chapters.

The image_picker input

The `image_picker` input type, as its name suggests, allows us to create an image picker selector field. The image picker selector allows us to upload new images, select the photo from a series of free photos available on Shopify, or use any of the photos previously included inside the **Files** section on the **Shopify** admin.

The `Files` section is home to all assets that we upload through admin or the theme editor, and we can find it inside **Shopify** by clicking on the **Settings** link in the bottom-left corner and then clicking on the **Files** link. In addition to acting as a database for all our assets, that is, the `Files` section, we can also upload assets directly into the `Files` section using the **Upload files** button in the top-right corner. Additionally, we can easily recover the direct URL path to any asset and use it.

In the following block of code, we can see the example of how we can use the `image_picker` input:

```
{
    "type": "image_picker",
    "id": "banner_image",
    "label": "Banner image"
}
```

Note that the `image_picker` type input value will always return an image object value or nil if the value is not defined:

Figure 6.18 – Example of the image_picker specialized input type

Since `image_picker` returns an `image` object, we can use the `img_tag` or `img_url` filters to generate the necessary image tag dynamically:

```
<img src="{{ settings.banner_image | img_url: "600x600" }}"
  alt="settings.banner_image.alt }}"/>
{{ settings.banner_image | img_tag: image_item.alt, "class1
  class2", "600x600" }}
```

For a detailed reminder regarding the use of `img_tag` and `img_url` filters, we can revisit the *HTML and URL filters* section from *Chapter 5, Diving into Liquid Core with Filters*.

The font_picker input

While learning about the `select` input type, we have mentioned that Shopify provides us with access to an extensive font library. The `font_picker` type input allows us to create a font picker selector field that we can use to select any font inside the Shopify font library:

```
{
    "type": "font_picker",
    "id": "body_font",
    "label": "Body font",
    "default": "helvetica_n4"
}
```

One additional change to the standard set of attributes is that the `default` attribute is now mandatory. We can find the possible font handle values for the `default` attribute at the following link, `https://shopify.dev/themes/architecture/settings/fonts#available-fonts`, by clicking on the **More info** button, located next to each font filename. The following screenshot shows us an example of the `font_picker` type input:

Figure 6.19 – Example of the font_picker specialized input type

Since the `default` value is mandatory, the `font_picker` value will always return a `font` object, allowing us to use the `font` filters and objects to manipulate the `font_picker` value to our needs.

Suppose we try to access the `font_picker` value using the regular approach, we would receive FontDrop as a result. To resolve this, we will include the `font_face` filter, which will generate the `@font-face` CSS:

```
<style>
    {{ settings.body_font | font_face }}
</style>
```

Declaring the `font_face` filter will automatically pull all the necessary information about the particular font and populate all the information inside `@font-face`:

```
<style>
  @font-face {
  font-family: Helvetica;
  font-weight: 400;
  font-style: normal;
  src: url("https://fonts.shopifycdn.com/helvetica/
  helvetica_n4.fe093fe9ca22a15354813c912484945a36b79146
  .woff2?&hmac=64c57d7fee8da8223a0d4856285068c02c248ef210ca
  e57dcd9c3e633375e8a4") format("woff2"),
    url("https://fonts.shopifycdn.com/helvetica
    /helvetica_n4.8bddb85c18a0094c427a9bf65dee963ad88de
    4e8.woff?&hmac=f74109e3105603c8a8cfbd8dec4e8a7e535
    72346fb96aacec203fc3881ddabf1") format("woff");
  }
</style>
```

With the `font_face` filter in place, we now have access to the selected font. However, with the current setup, we would have to hardcode the values to each `font-family`. So, let's learn how we can extract the `@font-face` attributes separately.

The first thing that we need to do is to create a variable to which we will save the font_ picker object value, after which we need to call the variable with the font_face filter:

```
<style>
  {% assign body_font = settings.body_font %}
  {{ body_font | font_face }}
</style>
```

With font_face declared, we now have easy access to @font-face from within the variable.

Let's say now that we wanted to modify particular attributes, such as font-weight and font-style, for specific elements. To achieve this type of functionality, we can use the font_modify filter, which accepts two properties, namely, the style property, which allows us to modify font-style, and weight, which we can use to modify the font-weight attribute:

```
<style>
  {% assign body_font = settings.body_font %}
  {{ body_font | font_face }}

  {% assign body_font_bold = body_font | font_modify:
    "weight", "bolder" %}
  {% assign body_font_italic = body_font | font_modify:
    "style", "italic" %}
  {% assign body_font_bold_italic = body_font_bold |
    font_modify: "style", "italic" %}
  {{ body_font_bold | font_face }}
  {{ body_font_italic | font_face }}
  {{ body_font_bold_italic | font_face }}
</style>
```

Note that we now have three different variables containing three different types of @ font-face. All that is left now is to extract the specific attributes that we need and assign them to style the content.

We can return the specific @font-face attribute values by using the font object's family, style, and weight attributes:

```
<style>
  {% assign body_font = settings.body_font %}
  {{ body_font | font_face }}

  {% assign body_font_bold = body_font | font_modify:
    "weight", "bolder" %}
  {% assign body_font_italic = body_font | font_modify:
    "style", "italic" %}
  {% assign body_font_bold_italic = body_font_bold |
    font_modify: "style", "italic" %}
  {{ body_font_bold | font_face }}
  {{ body_font_italic | font_face }}
  {{ body_font_bold_italic | font_face }}

  .body_bold {
    font-family: "{{ body_font_bold.family }}";
    font-style: "{{ body_font_bold.style }}";
    font-weight: "{{ body_font_bold.weight }}";
  }
  .body_italic {
    font-family: "{{ body_font_italic.family }}";
    font-style: "{{ body_font_italic.style }}";
    font-weight: "{{ body_font_italic.weight }}";
  }
</style>
```

We have now successfully learned how the font_picker input type works and, even more importantly, we have also learned how to use font objects and filters to output the font_picker values and make the font selection process entirely dynamic.

The only addition to the previous example that would make our code more functional is the inclusion of a fallback family if the selected font family cannot render for some reason. We can do this by introducing the `fallback_families` object, which will return a suggested fallback font family:

```
<style>
  {% assign body_font = settings.body_font %}
  {{ body_font | font_face }}


  {% assign body_font_bold = body_font | font_modify:
    "weight", "bolder" %}
  {% assign body_font_italic = body_font | font_modify:
    "style", "italic" %}
  {% assign body_font_bold_italic = body_font_bold |
    font_modify: "style", "italic" %}
  {{ body_font_bold | font_face }}
  {{ body_font_italic | font_face }}
  {{ body_font_bold_italic | font_face }}


  .body_bold {
    font-family: "{{ body_font_bold.family }}",
        "{{ body_font_bold.fallback_families }}";
    font-style: "{{ body_font_bold.style }}";
    font-weight: "{{ body_font_bold.weight }}";
  }
  .body_italic {
    font-family: "{{ body_font_italic.family }}",
        "{{ body_font_italic.fallback_families }}";
    font-style: "{{ body_font_italic.style }}";
    font-weight: "{{ body_font_italic.weight }}";
  }
</style>
```

For additional information on all the available font filters, we can refer to https://shopify.dev/api/liquid/filters/font-filters, and, for information on all the font objects available, we can refer to https://shopify.dev/api/liquid/objects/font.

The article input

The `article` type input provides us with a special article picker selector field. Through the article picker, we have access to all the available articles in the store:

```
{
    "type": "article",
    "id": "featured_article",
    "label": "Featured article"
}
```

Until recently, the `article` input type would always return a string handle of the article, which then needed to be used to recover the `article` object. However, since the Shopify Unite 2021 event, the `article` and other page-related inputs now return an object, making our work a lot easier.

Note that while the page-related input types will return an object from which we can extract any value we require, we will still sometimes find ourselves working on a theme where the theme uses the old method to retrieve the page-related input type object. For this reason, we will mention both approaches, since while we will not use the old approach, it is essential to know how it works if we ever need to work with it. In the following screenshot, we can see an example of the `article` input type:

SPECIALIZED INPUT SETTINGS   ▼

Featured article

Select article

Figure 6.20 – Example of the article specialized input type

Since the `article` input value returns an object value, we already have full access to the `article` object and can easily retrieve any attribute that we might need:

```
{% assign article = settings.featured_article %}
```

As we can see, now that the `article` input returns an object, accessing the object itself and retrieving the value of any attribute is almost effortless. Let's now look into the obsolete method to retrieve the `article` object using the article handle.

As mentioned previously, before the Shopify Unite 2021 event, the `article` input value returned a handle string. To access the `article` object, we will need to pluralize the name of the object we are trying to access, followed by a squared bracket `[]` notation, similar to what we did with the product handle in *Chapter 2, The Basic Flow of Liquid*:

```
{% assign article = articles[settings.featured_article] %}
```

We now have access to the `article` object and can easily output any type of article content to any part of the store. Note that while both approaches produce the same results, the second approach of recovering the page-related input object through its handle is now obsolete.

The blog input

The `blog` type input works similarly, as it provides us with a special blog picker selector field through which we have access to all the available articles in the store:

```
{
    "type": "blog",
    "id": "featured_blog",
    "label": "Featured blog"
}
```

Accessing the featured blog will always return a `blog` object or nil if the value has not yet been defined:

Figure 6.21 – Example of the blog specialized input type

Similar to the `article` object, we can directly access the `blog` object or retrieve it using squared brackets `[]` notation. The only difference is the changed `object` keyword:

```
{% assign blog = blogs[settings.featured_blog] %}
```

```
{% for article in blog %}
{% endfor %}
```

Once we have recovered the `blog` object, we can use the `for` tag to loop over all the articles inside the selected blog and render them correctly.

The collection input

The `collection` type input provides us with a special collection picker selector field, through which we will gain access to all the available products in the selected collection:

```
{
    "type": "collection",
    "id": "featured_collection",
    "label": "Featured Collection"
}
```

Similar to the previous settings type, the returned value will be nil if we have not yet selected the collection. Otherwise, the returned value will return an object value that we can use to retrieve any object attribute value:

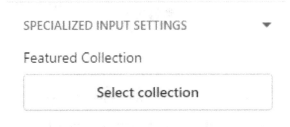

Figure 6.22 – Example of the collection specialized input type

Additionally, we can access the `collection` object directly or by pluralizing the object's name we are trying to access, followed by a squared bracket `[]` notation:

```
{% assign collection =
collections[settings.featured_collection] %}
```

With the `collection` object in place, we can now easily access all collection attributes.

The page input

The `page` type input provides us with another special selector field that allows us to access all of the pages previously created inside the admin **Page** section of our store:

```
{
    "type": "page",
    "id": "featured_page",
    "label": "Featured page"
}
```

The **Featured page** value, when accessed, will always return an object value or nil if the value is not yet defined:

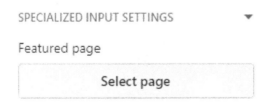

Figure 6.23 – Example of the page specialized input type

We can access the page object directly or through a squared bracket `[]` notation:

```
{% assign page = pages[settings.featured_page] %}
```

Using the declared page variable, we now have access to the `page` object, which we can use to further access any attribute for the specific page.

The product input

As its name suggests, `product` provides us with a product input selector, which we can use to gain access to the `product` object:

```
{
    "type": "product",
    "id": "featured_product",
    "label": "Featured product"
}
```

Similar to the previous settings, when accessed, the value will return nil if the value is not yet defined or we'll get an object value:

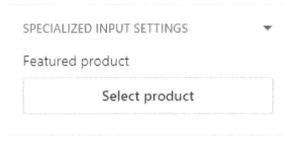

Figure 6.24 – Example of the product specialized input type

We can use the returned object value to access an attribute value, or we can use the `product` object's squared bracket `[]` notation to retrieve the object using the obsolete method:

```
{% assign product = products[settings.featured_product] %}
```

Through the product variable, we can now access all attributes within the selected `product` object.

So far, we have learned about all the different types of input settings, whether they are basic or specialized input types, how we can access them, what additional attributes we can include with each of them, and what type of value we can expect to receive from them. However, simply stacking a significant number of settings inside a single category can be overwhelming if there is no order to those settings.

Organizing the theme editor

In the previous section, we have learned about different types of configurable settings, which allows us to easily update its value using a series of basic or specialized input selectors, where now we will learn about a different set of settings, **sidebar settings**. The sidebar settings will allow us to divide each category's settings into other separate blocks.

The sidebar settings do not hold any value, nor can we configure them through the theme editor. Their only use is to provide us with additional information and help us organize different sets of input settings into separate blocks for more straightforward navigation.

Under the sidebar settings, we can use the following options:

- `header`
- `paragraph`

As opposed to the basic and specialized input settings, `header` and `paragraph` can only contain the following standard attributes:

- As its name indicates, the `type` attribute allows us to set the type of setting, whose value can be either a `header` or a `paragraph` setting. The `type` attribute is mandatory for both `header` and `paragraph`.

- While the sidebar settings cannot contain any value, we can use the `content` attribute to output certain information to the theme editor. The `content` attribute is also mandatory for both `header` and `paragraph`.

- The final attribute, `info`, allows us to provide additional information regarding the `header` type setting. Note that `info` is not a mandatory attribute and we can only use it with the `header` type setting.

Now that we have all the necessary information about the different types of sidebar settings and their attributes, let's look at each type of setting separately and learn how to use them.

The header type

The `header` type of setting, as its name suggests, allows us to create a header element and consequently group all of the setting input types inside a single block. Note that introducing the `header` type setting will automatically group all of the input type settings inside the category until it reaches another `header` element, or there are no more settings within the current category:

```
{
    "type": "header",
    "content": "Newsletter settings",
    "info": "Enabling the popup feature will automatically
        trigger a newsletter popup on page load."
}
```

As we can see, using the `header` type setting is relatively easy, where the benefit that we receive is quite significant, as now we can easily group related input settings. Additionally, by including the optional `info` attribute, we were able to include some additional information related to the specific block of settings, which we can see in the following screenshot:

Figure 6.25 – Example of the header sidebar type setting

We have now learned how to group related settings under the header name and include additional information about a particular block of settings. However, what if we needed to include an additional set of information to further describe a specific set of settings? To achieve this, we can use the following type of sidebar setting, `paragraph`.

The paragraph type

As mentioned, the `paragraph` type of sidebar setting allows us to include additional information, similar to using the `info` attribute with the header type setting:

```
{
    "type": "paragraph",
    "content": "It is recommended not to decrease the show
        after value below 3 seconds."
}
```

Note that including any additional attributes besides `type` and `content` with the `paragraph` type setting will result in an error. In the following screenshot, we can see an example using `paragraph` inside the theme editor:

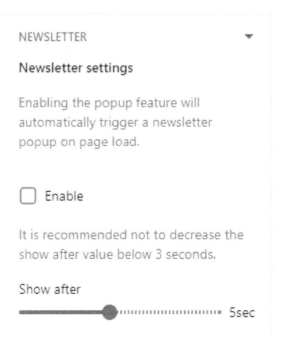

Figure 6.26 – Example of the paragraph sidebar type setting

While we cannot include additional attributes, we can use the `content` attribute to create a direct link to a specific page, providing some additional information. We can do this by including the necessary information by using the following format:

```
[link text](link url)
```

Including the direct link to a specific page will make our theme editor more concise. At the same time, it also allows us to include all the necessary information that someone using it might require:

```
{
    "type": "paragraph",
    "content": "The paragraph type setting allows us to
        include a direct link to any page. [Learn more](
            https://shopify.dev/themes/architecture/settings
            /sidebar-settings#header)"
}
```

Note that we can create a direct link to any page within both the `content` and the `info` attributes on any type of setting that accepts the `info` attribute.

We have now learned about every type of input and sidebar type of setting. However, while it is essential to know all the current settings, it is also essential to mention some of the now deprecated settings, but which we may still encounter in our everyday work.

Glancing at the deprecated settings

While the following settings are no longer supported, there is a high probability that we might encounter them on older themes that their store owner never updated. Since these settings are no longer supported, we will not go into that much detail on how they work, but we will provide some general guidelines for identifying them and what they do.

The font input

The `font` input type of setting allowed us to generate a shortlist of font files accessible on Shopify:

```
{
    "type": "font",
    "id": "body_font",
    "label": "Body font"
}
```

The introduction of the `font_picker` input type, which we have mentioned previously, made the `font` input type of setting obsolete.

The snippet input

As its name suggests, the `snippet` input type allowed us to select any snippet file that we created previously within our theme and execute its content in a specific position:

```
{
    "type": "snippet",
    "id": "featured_products",
    "label": "Featured products"
}
```

The introduction of `sections`, which we will learn about in the following chapter, has made the `snippet` input type of setting obsolete.

Summary

In the previous chapters, we have abstained from going over every specific option, consequently creating a list of options that we need to learn to keep everything concise and to the point. However, in this chapter, we have gone over each option and carefully explained how to use it, when, and what we can expect, since, on account of their importance, we will use the knowledge that we have learned in this chapter on a regular basis.

We have learned how to generate some of the most basic input types, which we can use to output various types of content and types of input, allowing us to create additional logic related to specific content. Additionally, we have also learned how to output specialized input types, allowing us to create complex features with an easy and configurable interface.

Lastly, we have learned how we can organize the JSON settings into separate blocks for better readability. The knowledge that we have attained through this chapter will be helpful and tested in the following chapter.

In the next chapter, we will learn about sections and blocks, and use them to create section/block-specific settings that merchants can use through the theme editor to create re-usable sections.

Questions

1. What are the two types of input settings?

2. What's the issue that will cause an error with the following piece of code?

    ```
    {
        "type": "text",
        "id": "header_announcement",
        "label": "Text",
    }
    ```

3. How can we include a custom font file within Shopify and use it throughout the theme editor?

4. What are the two issues that will prevent us from executing the following piece of code?

    ```
    {
        "type": "range",
        "id": "number_of_products",
        "min": 110,
        "max": 220,
        "step": 1,
        "unit": "pro",
        "label": "Number of products",
        "default": 235
    }.
    ```

7

Working with Static and Dynamic Sections

In the previous chapter, we familiarized ourselves with the different types of inputs, whether they're basic or specialized, and how we can use them to create global settings that we can easily configure through the theme editor.

In this chapter, we will not only get to use the previously mentioned inputs, but we will also learn how to create easily configurable and reusable sections that we can use to change the page or even templates' layouts easily.

We will learn about the following topics in this chapter:

- Static versus dynamic sections
- Working with the section schema
- Building with blocks
- Enhancing pages with JSON templates
- Exploring section-specific tags

By the time we complete this chapter, we will understand what sections are, when to use them, and how to create one. We will also learn about the difference between static and dynamic sections, and how we can configure them through the theme editor.

By learning about the section schema and the different attributes that we can use, we will also learn how to create reusable block modules within a section that we can use to repeat and receive different results. After familiarizing ourselves with sections and blocks, we will learn how to improve these concepts by learning about JSON templates and built-in metafields. Lastly, we will learn about the different types of specialized tags that we can use within section files, which will help us create reusable and dynamic modules.

Technical requirements

While we will explain each topic and have it presented with the accompanying graphics, we will need an internet connection to follow the steps outlined in this chapter, considering that Shopify is a hosted service.

The code for this chapter is available on GitHub: `https://github.com/ PacktPublishing/Shopify-Theme-Customization-with-Liquid/tree/ main/Chapter07`.

The Code in Action video for the chapter can be found here: `https://bit.ly/3hQzhVg`

Static versus dynamic sections

In *Chapter 1*, *Getting Started with Shopify*, while discussing the theme structure, we briefly mentioned section files, *but what exactly are sections?*

Besides being the name of one of the directories in the theme file, a **section** is a type of file that allows us to create reusable modules that we can customize using the theme editor, as we learned previously. However, as opposed to the global settings that we have learned about, the major difference is that the JSON settings for sections are defined inside each section file and are section-specific.

Section-specific settings allow us to reuse the same section module multiple times on the page and select a different set of options for each occurrence, making it a pretty powerful feature. For example, we can create a featured collections section and repeat it three times to display three products from three separate collections.

Before we proceed, let's navigate to the theme editor by clicking the **Customize** button from the code editor and see the section files in action. Clicking the **Customize** button will automatically open the theme editor and position us on the home page, with the page preview on the right-hand side and a sidebar on the left. Within the sidebar, we can see several sections that are available on this specific page:

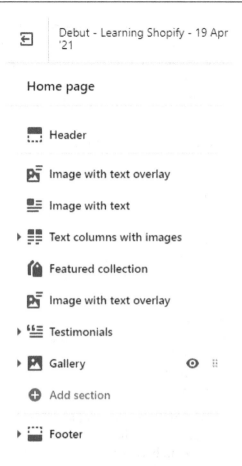

Figure 7.1 – Example of page sections within the theme editor

At first glance, we can see that the **Header** and **Footer** sections are separated from the rest of the sections by a thin border. This tells us that the **Header** and **Footer** sections are static sections whose positions we cannot change from within the theme editor.

Similarly, as with snippet files, **static sections** are sections that we can manually insert within the theme files, which we can do using the `section` theme tag:

```
{% section "name-of-the-file" %}
```

By including the `section` tag, we will automatically render the content of the section to the position where we have included the tag and be able to customize it further using its JSON settings. Note that since this is a static section that we have manually included within the theme files, any settings that we apply to this specific section through the theme editor will be visible across all the pages where we have included this section.

For example, the theme developers have included the header and footer section files inside the theme.liquid layout file, which is our theme master file, meaning that the header and footer sections will be visible on any page on our store. On the other hand, if we were to navigate to a different page within our theme editor, we would see a header, a footer, and an entirely different set of sections between those two than those we previously saw on the home page.

By including the header and footer sections inside the theme.liquid file, we have made them visible across the entire theme on any page. However, we will not include all the sections within the theme.liquid file as they are not needed. Instead, we will include certain template-specific sections inside their respective template files so that they can only be accessed when visiting pages with the specific template assigned.

Let's say that we wanted to include the related products section on the product pages. Here, we would navigate to the product.liquid template and include the section tag at any position, which would make the related products section visible on any product that uses this particular template.

As we recall from *Chapter 4, Diving into Liquid Core with Objects*, in *The Please apply P-Italics here content_for_layout object* section, we mentioned that the content_for_layout object allows us to connect the other templates to the theme.liquid file by loading the dynamically generated content, including the section files from other templates, into the theme.liquid file.

Since content_for_layout outputs the content of all other templates, by placing it between the header and footer sections, we have ensured that we will position all the section files from other template files in between the header and footer sections. Let's look at the theme.liquid layout file; we will notice that we positioned content_for_layout between the header and footer sections, as depicted in the following block of code:

```
{% section 'header' %}
  <div class="page-container drawer-page-content"
    id="PageContainer">
    <main class="main-content js-focus-hidden"
        id="MainContent" role="main" tabindex="-1">
      {{ content_for_layout }}
    </main>
    {% section 'footer' %}
```

Note that including the same section file across multiple templates will result in the same content being visible across all templates. By configuring the static section from the theme editor, we can save the selected data inside the settings_data.json file, which will return the exact data for any occurrence of the same section.

If we need to repeat the static section multiple times with different content, we would need to create a new section file using a different name:

```
{% section "related-product-1" %}
{% section "related-product-2" %}
```

So far, we have seen what static sections are and how to use them to create a template-based configurable layout. But we also have access to dynamic sections, which we do not have to include every time we want to reposition a section manually.

As the name suggests, **dynamic sections** are a set of sections that we can add, remove, reposition, or repeat any number of times with different content, without touching a single line of code, all from the theme editor. Let's return to the home page within the theme editor and look at some of the existing dynamic sections:

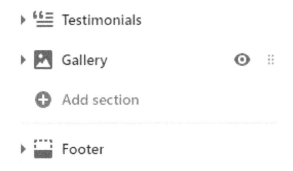

Figure 7.2 – Example of page sections within the theme editor

We can easily distinguish the static section from the dynamic section by simply hovering over the section itself. We will notice two icons to the right of the dynamic section's name – an *eye* icon and *six dots* – by hovering over it. The eye icon will allow us to toggle the section's visibility on and off. By clicking and holding onto the six dots icon, we can reposition the section by moving it above or below the other sections.

Additionally, at the bottom of the dynamic sections, we will notice an **Add section** button, which, once clicked, will show a dropdown that we can use to include any number of sections that exist in our store by simply clicking on them. Any section that's added to the theme via the **Add section** button will allow us to include different content for any occurrence, and we can repeat this as many times as we want.

Until recently, the home page was the only place where we could use dynamic sections. For all other templates, we had to rely on using static sections. However, since the Shopify Unite 2021 event, Shopify has introduced JSON type templates, which we briefly mentioned in *Chapter 1, Getting Started with Shopify*. JSON templates, which we will learn more about later in this chapter, add dynamic sections to any page and bring the store's entire functionality to a whole new level.

Now that we have learned about what sections are and how we can use them, it is time to learn how to create them.

Working with the section schema

In the previous chapter, we learned how to use JSON to create global settings, which has a similar format to the JSON for section files. However, sectional JSON comes with a few distinct differences.

The first major difference is that we need to define JSON inside the section file instead of the `settings_schema.json` file. To do this, we will need to introduce the `schema` tag:

```
{% schema %}{% endschema %}
```

The `schema` tag is a Liquid tag that does not have any output on its own. It simply allows us to write JSON code inside the section files. Note that each section file can only contain a single `schema` tag, which must stand on its own. It cannot be nested inside any kind of other Liquid tags.

Once we have the `schema` tag in place, we are ready to familiarize ourselves with the types of attributes that we can use within the `schema` tags.

The name attribute

As its name suggests, the name attribute allows us to set the section name, which we will use to identify the section inside the theme editor later:

```
{% schema %}
{
  "name": "Announcement bar"
}
{% endschema %}
```

With that, we have learned how to define the name of the section file, *but what if we wanted to create an international store where we can easily translate not only the store content, but also the store setting labels, inside the theme editor?*

We can easily translate most of the attributes inside the schema tag by including the translation keys in the name attribute value:

```
{% schema %}
{
  "name": {
    "cs": "Panel oznámení",
    "da": "Meddelelseslinje",
    "de": "Ankündigungsbereich",
    "en": "Announcement bar"
  }
}
{% endschema %}
```

The first value represents the name of the file inside the Locales directory, whereas the second one represents the translated value. By introducing the translating keys, we have ensured that as soon as we change our store language, the name value will automatically adjust and present the value of the currently selected language file.

Note that we can include the translation keys for different kinds of attributes, including some of the input settings that we learned about in the previous chapter. The attributes that we can use the translation keys with include name, info, label, group, placeholder, unit, content, and category.

Note that name is a mandatory attribute. However, the name attribute does not have to be unique compared to the other sections, so we should pay attention when creating new sections to avoid confusion.

The class attribute

`class` is a simple attribute that allows us to add additional classes to the `div` element, which wraps around the section content:

```
{% schema %}
{
  "name": {
    "cs": "Panel oznámení",
    "da": "Meddelelseslinje",
    "de": "Ankündigungsbereich",
    "en": "Announcement bar"
  },
  "class": "homepage-section desktop-only"
}
{% endschema %}
```

Through this optional attribute, we can easily include any number of classes in the parent element wrapping around our section, as shown in the following code block:

```
<div id="shopify-section-[id]" class="homepage-section
  desktop-only">
</div>
```

Note that while we can include any number of classes, we cannot dynamically modify them since the `class` attribute only accepts string values.

The settings attribute

Using the `settings` attribute, we can create section-specific settings, which allows us to configure the section using the theme editor:

```
{% schema %}
{
  "name": {
    "cs": "Panel oznámení",
    "da": "Meddelelseslinje",
    "de": "Ankündigungsbereich",
    "en": "Announcement bar"
  },
```

```
    "class": "homepage-section desktop-only",
    "settings": [
    ]
  }
{% endschema %}
```

Once we have defined the `settings` attribute, we can start including any input settings, whether they're basic or specialized, that we have previously learned about to create the flow that we need. Since we are working on creating an announcement bar, we can include the following inputs:

```
{% schema %}
{
  "name": {
    "cs": "Panel oznámení",
    "da": "Meddelelseslinje",
    "de": "Ankündigungsbereich",
    "en": "Announcement bar"
  },
  "class": "homepage-section desktop-only",
  "settings": [
    {
      "type": "text",
      "id": "announcement-text",
      "label": "Text"
    },
    {
      "type": "color",
      "id": "announcement-text-color",
      "label": "Text color",
      "default": "#000000"
    }
  ]
}
{% endschema %}
```

With the introduction of `settings` and input settings, we have successfully created the first section, whose text and text color values can be adjusted within the theme editor. If we were to try and include this section as a static section, the two input settings that we created previously would immediately be visible under the respective section:

```
{% section "section-file-name" %}
```

However, if we were to navigate to the home page and include the announcement section via the **Add section** button, we would not find it. We are still missing one attribute, which we need to create a section that we can dynamically add to the home page.

The presets attribute

The `presets` attribute allows us to define the default configuration of a section, which makes the section accessible from the **Add section** dropdown. The `presets` attribute can contain the following attributes:

- The `name` attribute is mandatory and will define how the section will appear under the **Add section** dropdown.

- The `category` attribute is not mandatory. We can use it to group different sections under a single category for more straightforward navigation.

Note that it is advisable to use a unique name for the `presets` section to avoid confusion, even though Shopify does not strictly require it. Otherwise, we might end up with multiple sections with similar names:

```
{% schema %}
{
  "name": {
    "cs": "Panel oznámení",
    "da": "Meddelelseslinje",
    "de": "Ankündigungsbereich",
    "en": "Announcement bar"
  },
  "class": "homepage-section desktop-only",
  "settings": [
    {
      "type": "text",
      "id": "announcement-text",
      "label": "Text"
```

```
    },
    {
        "type": "color",
        "id": "announcement-text-color",
        "label": "Text color",
        "default": "#000000"
    }
  ],
  "presets": [
    {
        "name": "Announcement bar",
        "category": "Text"
    }
  ]
}
{% endschema %}
```

Note that we should only include the `presets` attribute for dynamic sections. If we plan on using it as a static section, we should remove the `presets` attribute.

If we were to open the theme editor now and click on the **Add section** button on the home page, we would see the **Announcement Bar** section, with the text and color settings that we defined earlier.

Accessing the section-specific JSON input values is relatively similar to how we access the settings inside the `settings_schema.json` file. The only difference is that the only way to access the section's `settings` object is through the `section` object:

```
{{ section.settings.announcement-text }}
{{ section.settings.announcement-text-color }}
```

For additional information on the `section` object, please refer to `https://shopify.dev/api/liquid/objects/section`.

So far, we have learned what sections are, the differences between static and dynamic sections, and how to use them. However, looking over the announcement section project that we have worked on, it is clear that the entire section is pretty basic as it only allows us to create a single announcement.

We can include a few more text input options that we can use to create multiple announcements, but that would require us to manually edit the JSON code every time we need to include an additional announcement. *What if we're looking to create a section that would allow us to add any number of announcements, without modifying the JSON code every time we need to include an additional announcement?* For this, we can use the `blocks` attribute.

Building with blocks

The `blocks` attribute is one of the most potent tools in Shopify. By using them, we can create modules that we can reuse any number of times and reorder the section content from within the theme editor. At this point, this might be confusing as it sounds similar to what we just learned about regarding dynamic sections. However, the key difference is that `blocks` sections allow us to reorder the content within the section, not the sections themselves, which allows us to create more complex features.

Additionally, we can combine the `blocks` attributes with static sections to create a dynamic section functionality that's similar to what we currently have on the home page, and then include it on any page. However, instead of rearranging sections, we will be rearranging the blocks.

The `blocks` attribute allows us to create different types of blocks using the object format, under which each object type will act as a unique module. Here, we can include a different set of input settings options under each of those modules.

Here are the steps for creating and using a `blocks` module:

1. Let's learn how to modify the previously created announcement bar section by introducing the `blocks` attribute. Additionally, we will remove certain features, such as translation keys and the `class` attributes, for the code to remain concise and readable:

```
{% schema %}
{
  "name": "Announcement Bar",
  "settings": [
    {
      "type": "text",
      "id": "announcement-text",
      "label": "Text"
    },
```

```
        {
            "type": "color",
            "id": "announcement-text-color",
            "label": "Text color",
            "default": "#000000"
        }
    ],
    "blocks": [
    ],
    "presets": [
        {
            "name": "Announcement bar",
            "category": "Text"
        }
    ]
}
{% endschema %}
```

Like creating a section, the `blocks` attribute has its own set of attributes that we need to use to create different modules:

- The `name` attribute allows us to set the name of the `blocks` module and decide how the block will appear in the theme editor. The `name` attribute is mandatory.

- The `type` attribute is a mandatory attribute that accepts a string value where we can define the block type. Note that the `type` attribute does not have a predefined set of values. Instead, we can use any string value to define the block type.

- Using the `limit` attribute, we can limit how many times we can repeat a particular block type. The `limit` attribute is optional and only accepts a `number` type value.

- The `settings` attribute allows us to include `blocks` module-specific settings. The `settings` attribute is optional.

Note that the `name` and `type` attributes of the `blocks` module need to remain unique within the section, whereas the `id` attribute needs to remain unique within the `blocks` modules. Otherwise, we will end up with invalid JSON code.

2. Now, let's learn how to include the previously mentioned `blocks` module attributes inside the section schema. Note that we will only show the code that's inside the `blocks` attribute to keep the code concise and to the point:

```
"blocks": [
    {
        "name": "Announcement",
        "type": "announcement",
        "limit": 3,
        "settings": [
        ]
    }
]
```

In the previous example, we created a `blocks` module with the `name` attribute set to `Announcement`, the `type` attribute set to `announcement`, and the `limit` attribute set to `3`, which limits the `blocks` module to a maximum of 3 repetitions.

3. Now that we have all the necessary attributes in place, all we need to do is populate the `settings` attribute with the necessary input settings. Since we have already created the text and text color input settings, we can simply migrate the text input setting inside the `blocks` module:

```
"blocks": [
    {
        "name": "Announcement",
        "type": "announcement",
        "limit": 3,
        "settings": [
            {
                "type": "text",
                "id": "announcement-text",
                "label": "Text"
            }
        ]
    }
]
```

With the text input inside the `blocks` module, we can dynamically repeat the entire block up to three times, consequently creating three separate announcements.

4. Let's see what this looks like by navigating to the theme editor and clicking on the **Add section** button to include the **Announcement Bar** section on the home page. The first thing that we will notice is the **Text color** option, which we have left inside the section `settings` object:

‹   Announcement Bar

Text color

None

Figure 7.3 – Example of a section inside the theme editor

Since we have only migrated the text input and left the text color input inside the section `settings` object, we can use it to style the text color of all the blocks at once. Let's click on the arrow to the left of the **Announcement Bar** title to return to the previous section and see our `blocks` module in action.

We will notice that besides the **Add section** button, we also have an **Add Announcement** button under the **Announcement Bar** section that we created:

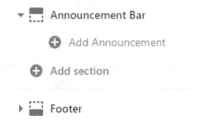

▼ Announcement Bar

   ⊕ Add Announcement

⊕ Add section

▸ Footer

Figure 7.4 – Example of a section block module

5. Clicking **Announcement Bar** will return us to the previous page, where we can configure the settings inside the section `settings` object, whereby clicking on the **Add Announcement** button will automatically include a single `blocks` module and immediately position us inside the block. Here, we can define the announcement text input.

6.   After defining all the inputs inside the `blocks` module's object settings, we can click on the arrow on the left-hand side of the **Announcement** title to return to the previous page, where we can include additional `blocks` modules. However, note that since we have introduced a `limit` attribute with a value of 3, we can only repeat the **Announcement** block up to 3 times, as shown here:

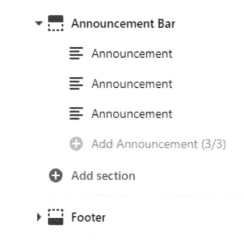

Figure 7.5 – Example of a limited number of section block modules

Once we have included the maximum number of modules, as per the defined value, the **Add Announcement** button will turn gray, signaling that we have reached the maximum number of `blocks` modules. Similarly, as with the dynamic sections, hovering over the `blocks` modules will reveal two icons. These will allow us to hide the currently selected block or rearrange the block's order using the drag-and-drop feature.

Now that we have learned how to create and use a `blocks` module, we need to learn how to output the values of the `blocks` module's input settings:

1.   We can access the `blocks` module's object settings similar to how we accessed the `section` object: we will need to use a `section` object in combination with the `blocks` attribute. This combination of the `section` object and the `blocks` attribute will return an array of sections blocks that we can access using a simple `for` loop:

```
{% for block in section.blocks %}
{% endfor %}
```

2. Once we have created the `for` loop, the only thing left to do is output the values of each block, as we did with the sections. The only difference is that this time, we will use the variable that we defined in the `for` loop instead of using the `sections` keyword:

```
{% for block in section.blocks %}
    {{ block.settings.announcement-text }}
{% endfor %}
```

So far, we have learned how to create the entire section schema, through which we can create static and dynamic sections and build `blocks` modules. We have also learned how to output the values of both the `sections` and `blocks` modules' input settings. However, in the previous example, we only had one type of block; *what if we had multiple block types?*

The true power of section blocks is that we can create multiple `blocks` module types inside a single `section` element, which we can do by simply creating multiple section `blocks` objects with different type values:

```
"blocks": [
  {
    "name": "Product",
    "type": "product",
    "settings": [
      {
        "type": "product",
        "id": "featured-product",
        "label": "Product"
      }
    ]
  },
  {
    "name": "Collection",
    "type": "collection",
    "settings": [
      {
        "type": "collection",
        "id": "featured-collection",
        "label": "collection"
```

```
        }
    ]
  }
]
```

Note that all of the `name` and `type` attribute values need to be unique across the entire section, whereas the `id` attributes of the input settings only need to be unique inside a single block.

As we can see, creating multiple `blocks` module elements is relatively simple. While we have only created two simple `blocks` modules in our example, we can create any number of `blocks` modules that we can rearrange through the theme editor to create complex layouts:

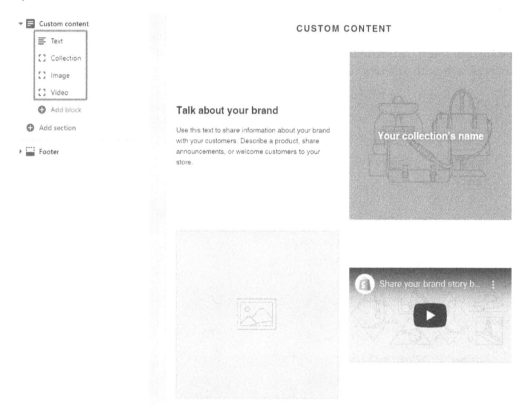

Figure 7.6 – Example of multiple blocks modules inside a section

As we can see from our example, we have four different types of blocks (**Text**, **Collection**, **Image**, and **Video**) that we can use to create a layout that the store owners can easily configure, without having to modify the code themselves.

Previously, we learned how to loop over the array of blocks by using the `for` tags that `section.blocks` returns. However, if we have more than one `blocks` module type, we will need to introduce an additional step.

Suppose that, as in the preceding example, we had a section with four types of blocks. *How would we recognize the block types and know what layout we should output for which block?* To solve this problem, we can use the `block` object paired with the `type` attribute, which will allow us to recover the block type value.

For additional information on the `block` object, please refer to `https://shopify.dev/api/liquid/objects/block`.

Once we have identified each block type, we can use an `if` statement or `case/when` control flow tags to execute the correct code for the respective value. Suppose we need to remind ourselves how to use statements or `case/when` control tags; we can revisit *Chapter 3*, *Diving into Liquid Code with Tags*, and consult the *Controlling the flow of Liquid* section for this, where we outlined the process of using control flow tags:

```
{% for block in section.blocks %}
  {% case block.type %}
    {% when "collection" %}
      {% render "block-collection ", collection: block %}
    {% when "image" %}
      {% render "block-image", image: block %}
    {% when "video" %}
      {% render "block-video", video: block %}
  {% endcase %}
{% endfor %}
```

In the previous example, we used the `case/when` control flow tags in combination with `block.type` to identify the types of blocks that we are currently looping over. After identifying the block type, we rendered the snippet file containing the correct layout for the respective block.

Note that we have passed the `block` object to each snippet. As you may recall, the snippet files are the only type of files that can access the variables defined in the parent directory. However, even the snippet files cannot automatically access these variables. Instead, we need to pass the values as parameters.

Suppose we need to remind ourselves how to pass values to snippet files; we can revisit *Chapter 3*, *Diving into Liquid Code with Tags*, and consult the *The render tag* section, located under the *Theme tags* section, where we outlined the process of working with snippet files.

Besides learning how to create `blocks` modules, we learned about the `limit` attribute, which we can use to limit how many times we can repeat a particular type of `blocks` module. This worked well when we had a single type block. *However, what if we wanted to create a limitation so that we can have a maximum number of any combination of blocks inside a section?*

The max_blocks attribute

The `max_blocks` attribute, similar to the `limit` attribute, allows us to limit how many `blocks` modules we can create inside a section. However, they have one significant difference: the `limit` attribute only allows us to limit the number of times we can repeat a particular block type, whereas the `max_blocks` attribute allows us to limit a specific section to a maximum number of any type of block.

Note that the `max_blocks` attribute is optional and only accepts number data as its value:

```
{% schema %}
{
  "name": "Footer",
  "max_blocks": 5,
  "settings": [
  ],
  "blocks": [
  ]
}
{% endschema %}
```

The most common use of the `max_blocks` attribute is inside the footer section. With it, the store owner can easily rearrange the blocks in any order, whether by repeating a single block five times or by using five different blocks, all while ensuring that the layout of the entire section maintains the proper flow.

Previously, we learned how to create `blocks` modules, as well as how to identify different types of block types and access each block object accordingly. Now that we have learned how we can use `max_blocks` to limit the number of products inside a section, we have all the necessary knowledge to build any type of `blocks` module.

As we saw, blocks are quite powerful and allow us to create anything from basic text features to complex layout features, intertwined with different types of blocks that store owners can use to tell the story of their products.

With that, we have learned about the differences between static sections, which we can manually embed on any page, and dynamic sections, which allow us to add any number of sections to the home page dynamically. However, as we mentioned at the beginning of this chapter, Shopify has recently provided us with the means to include sections on any page through JSON templates dynamically. By using JSON templates, we can combine the static and dynamic sections into a new feature that we can control from within the theme editor.

Enhancing pages with JSON templates

In *Chapter 1*, *Getting Started with Shopify*, while discussing the theme structure, we briefly mentioned JSON templates, *but what exactly are they?*

The `.json` type templates generally have the same purpose as their counterpart `.liquid` templates, as they both allow us to create and manage the look of multiple pages through a single template. However, the significant difference between the two is that while the `.liquid` type template serves only as a markup file, the `.json` file serves as a data type file, which allows us to easily add, remove, or rearrange the sections on any page, similar to what we can do on the home page.

The `.json` type templates also share similarities with the `Section` directory files, where we need to include a valid schema setting inside the section file. The `.json` type template needs to be a valid `.json` file, with the JSON code defined inside the template. While we can create any number of `.json` type files, the template files' names must be unique in the `.liquid` or `.json` file. For example, if we create a `product.json` file template, we cannot create a `product.liquid` file as well.

Additionally, JSON files have one limitation. We can render a maximum of 20 sections per template with up to 16 blocks per section, which is a reasonably high number, but we should probably rethink our page layout if we ever reach this limit.

Now that we have some general knowledge of what JSON type templates are and how they work, let's learn how to create our first JSON type file.

Building a JSON template structure

Instead of simply listing the necessary attributes for creating the file, we will learn how to create a JSON template by migrating the current `product.liquid` type template into a JSON template.

Let's begin by creating a JSON-type template by opening the `Templates` directory and clicking the **Add a new template** button. A popup will appear, where we can select which page we are creating the template name for, the template type, and the filename. For our purposes, we'll select the **product** page from the page selection dropdown and select the **JSON** radio button for **Template type**. As for **File name**, we can leave it as the default; that is, `alternate`:

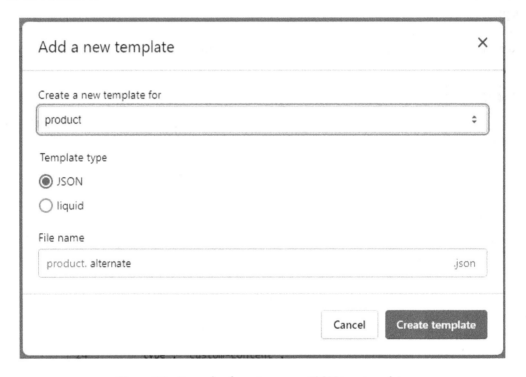

Figure 7.7 – Example of creating a new JSON type template

Once we have created a new JSON file, Shopify will automatically redirect us to the newly created file. We will see an almost empty file containing only two attributes. So, let's start creating the necessary attributes and begin our migration process.

As we mentioned previously, a JSON template must be a valid JSON file whose root is an object that can contain some of the following attributes:

- The name attribute is a mandatory string type attribute where, as its name suggests, we can define the template's name:

```
{
    "name": "JSON product template"
}
```

- The layout attribute is an optional attribute that accepts two types of values, a string or false, depending on what we are trying to achieve. The string value should represent the name of the layout file, without the .liquid extension that we would like to use with this specific template. If you need to learn more about layout files, please revisit *Chapter 3, Diving into Liquid Core with Tags*, where we learned about layout tags in the *Theme tags* sections. Note that if we do not include the layout attribute, Shopify will default to the theme.liquid layout. However, for learning purposes, we will manually add and select the theme layout:

```
{
    "name": "JSON product template",
    "layout": "theme"
}
```

- The wrapper attribute is a string type attribute that allows us to select the type of HTML wrapper we would like to include around each section inside the template. We can use the following HTML tags here:

  - div
  - main
  - section

  Note that besides selecting the HTML wrapper's type, we can also include any additional attributes that we may need, such as class, id, or data. Note that the wrapper attribute is entirely optional:

```
{
    "name": "JSON product template",
    "layout": "theme",
    "wrapper": "div.product-wrapper[data-type=product]"
}
```

- The `sections` attribute is a mandatory object type attribute that uses the names of the sections as keys and the `section` data as values. Inside the `sections` object, we can define which sections we would like to include inside our page:

```
{
    "name": "JSON product template",
    "layout": "theme",
    "wrapper": "div.product-wrapper[data-type=product]",
    "sections": {
    }
}
```

We might be wondering why we need to manually select which sections we would like to include since the whole point of dynamic sections is to avoid this.

The `sections` object allows us to include a mandatory static section, similar to our main page content, that is more page-specific. First, we will need to set a unique ID representing each static section that we are looking to include:

```
{
    "name": "JSON product template",
    "layout": "theme",
    "wrapper": "div.product-wrapper[data-type=product]",
    "sections": {
        "main-block": {
        },
        "different-block": {
        }
    }
}
```

Considering that the `sections` attribute uses the same format as the `section` attribute that we learned about previously, we will also need to include some additional attributes inside the `sections` object. For example, we will need to include the `type` attribute, whose value should have the name of the section we are looking to include, and if needed, the `settings` or `blocks` attribute:

```
{
    "name": "JSON product template",
    "layout": "theme",
    "wrapper": "div.product-wrapper[data-type=product]",
```

```json
    "sections": {
      "main-block": {
        "type": "name-of-the-section",
      },
      "different-block": {
        "type": "name-of-another-section"
      }
    }
  }
```

However, note that the `settings` or `blocks` attribute should not contain the input type settings. Instead, we should set the key value to the ID of the existing input inside each section, as well as the value of the `settings` input value that we wish to set:

```json
{
  "name": "JSON product template",
  "layout": "theme",
  "wrapper": "div.product-wrapper[data-type=product]",
  "sections": {
    "main-block": {
      "type": "name-of-the-section",
      "settings": {
        "show_discount": true,
        "gallery_type": "slider"
      },
      "blocks": {
      }
    },
    "different-block": {
      "type": "name-of-another-section"
    }
  }
}
```

Now, let's update the previous example and include the actual static sections we are currently using inside our default Liquid-type product template:

```
{
    "name": "JSON product template",
    "layout": "theme",
    "wrapper": "div.product-wrapper[data-type=product]",
    "sections": {
      "main": {
        "type": "product-template"
      },
      "recommendations": {
        "type": "product-recommendations"
      }
    }
}
```

Note that the names of the sections that we have used (main and recommendations) are not preset values and that we can change their names to any values we like.

When dealing with the sections object, there are three crucial things that we need to keep in mind:

- We need to have at least one block set inside the sections object.
- All block names must be unique across the entire sections object.
- The value of the type attribute needs to match the name of the section we are looking to include.

By including the main and recommendations sections in our new JSON template, we have ensured that these two sections will always be visible inside the theme editor when previewing the page with the specific JSON type template assigned. However, while these sections are considered static, we can hide them and rearrange them, similar to dynamic sections.

- The last attribute is the `order` attribute, which is a mandatory attribute. The `order` attribute is an array-type attribute. Here, we can include the IDs of the `sections` blocks that we previously set inside the `sections` object and arrange them:

```
{
    "name": "JSON product template",
    "layout": "theme",
    "wrapper": "div.product-wrapper[data-type=product]",
    "sections": {
      "main": {
        "type": "product-template"
      },
      "recommendations": {
        "type": "product-recommendations"
      }
    },
    "order": [
      "main",
      "recommendations"
    ]
}
```

And with the `order` attribute in place, our JSON template is ready! Let's navigate to the theme editor and test it out.

To test the new product template, follow these steps:

1. Navigate to **Products** from the admin dashboard.

2. Click on any product and then select the new template name from the **Template suffix** drop-down menu, which is located under the **Theme templates** area.

> **Important note:**
> The **Template suffix** drop-down menu can only read values from the current live theme. What this means is that the newly created template file will not be visible in our admin dashboard until we either publish or duplicate them live, or until we create the same template file within our currently live team. If we opt for the latter choice, note that we need to create the file with the same name; we do not have to make any changes to the file's content.

However, with this new type of template, we also have a new way to preview templates. Note that the following method only allows us to preview the template. We still need to assign the template by navigating to the admin dashboard's product page and selecting the template from inside the dropdown:

1. Let's begin by navigating to the theme editor, clicking on the dropdown in the middle of the screen, and selecting the **product** option. This will show you the templates we currently have under the theme that we are currently working on. Let's select the new JSON type template that we created by clicking on it:

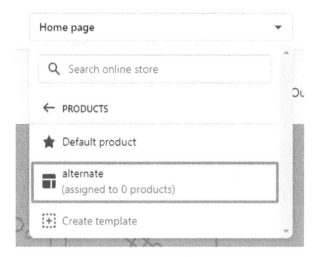

Figure 7.8 – Example of selecting a template through the theme editor

2. Clicking on the **alternate** template will automatically redirect us to a random product, allowing us to preview the template that we have selected.

> **Important note:**
>
> Similar to the dynamic sections on the home page, only sections with preset attributes present inside the sections schema will be visible under the **Add section** dropdown.

If we have done everything correctly, we should see two sections and the **Add section** button in the left sidebar:

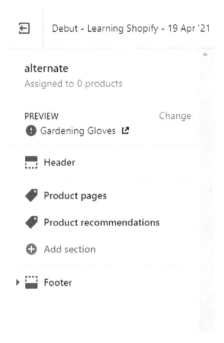

Figure 7.9 – Example of previewing the JSON type template inside the theme editor

3. Now, try and click on any of the two sections, update any settings, or add a new section through the **Add section** button and press the **Save** button in the top-right corner. Next, return to the code editor, close the new .json type template we created by clicking on the **X** next to its name, and then open it again.

    We will notice that **Shopify** has updated the template file and that it now contains all the settings and their values.

    Since we can find all of the settings and their values inside the template file, if we assign the JSON template file to multiple products, they will use the layout with the same settings. If we wanted to create an additional JSON template that would allow us to create a different layout, we could manually create a new JSON template and copy the code or do this through the theme editor.

4. Let's return to the theme editor, click on the dropdown in the middle of the screen, and select the **product** option. But this time, we will click on the **Create template** button, which will cause the following popup to appear:

Figure 7.10 – Example of creating a new JSON type template through the theme editor

As we can see, inside the popup, we can easily set a new template name and select the JSON template whose content we should copy.

With that, we have learned how to create any number of templates for any number of pages, *but what if we wanted to use the same layout for multiple pages and only use different content for each page?* This is where metafields come to the rescue!

Upgrading a JSON template with metafields

As you may recall from *Chapter 4, Diving into Liquid Core with Objects*, while going through the *Improving the workflow with metafields* section, we mentioned that Shopify has introduced a feature that allows us to use metafields without having to rely on third-party apps.

Besides allowing us to create metafields within the dashboard, Shopify has provided us with a whole new set of different types of metafields that we can create:

1. Start by navigating to the **Settings** options inside our dashboard, located at the bottom-left corner of our screen, and click on the **Metafields** options.

2. Once inside, we will see that we have no metafields set and that we can only use products and variants metafields; the others are still pending. Let's proceed by clicking on the **Products** metafield link.

3. Considering that we have no metafields currently set, we can immediately click on the **Add definition** button. This will redirect us to a page where we can create a metafield definition. While most of the fields should be familiar from when you learned about metafields in *Chapter 4, Diving into Liquid Core with Objects*, we now have a new field. Let's click on the **Select content type** field, which will show us a dropdown containing all the available types of metafields we can create:

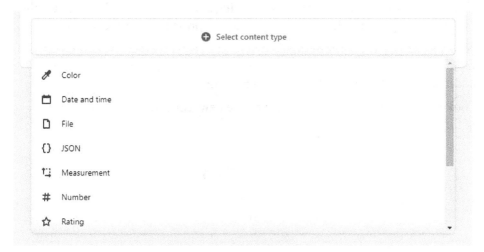

Figure 7.11 – Example of the available metafields type fields

Note that each type of metafield will create a selector inside the dashboard of our product page. We can select the text input, which will automatically show us an additional set of settings. However, we won't be changing those now.

4. After selecting the text type of the metafield, all we need to do is set its name. For our purposes, we will select one of the predefined metafield values by clicking on the **Name** field and selecting **Product subtitle**, which will automatically fill in all the other fields. Note that we will need to press the **Save** button to save the newly created metafield:

Figure 7.12 – Example of the metafields inside the product dashboard

5. Now that we have created a product metafield, let's click on the **Products** link located in the top-left corner of our admin dashboard and click on any product of our choosing. If we scroll down to the bottom of the page, we will notice that the metafield we created is now part of our product pages.

6. As we can see, by simply creating a product metafield definition, we have automatically added the same metafield to every product, allowing us to update the metafield's values a lot easier than when using a third-party app. Let's update the metafield's value by adding any string value and pressing the **Save** button.

7. Now, return to the theme editor, navigate to the JSON template, and find any text input type from the sections that we have available. For our purposes, we have added a new section named Image with text.

8. Looking through this section, we will notice new icons next to specific types of fields:

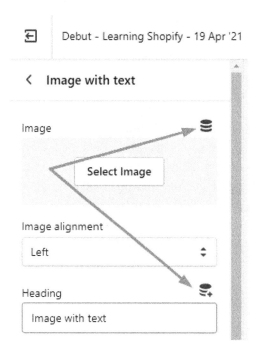

Figure 7.13 – Example of the metafields icon inside the theme editor

Clicking on the bottom icon next to the **Heading** text input will promptly show us a list of all the metafields that we can pull from this particular product.

9. Select the **Product subtitle** metafield that we selected previously and click the **Insert** button. This will automatically add the value of the selected metafield and output it in the input field that we have selected:

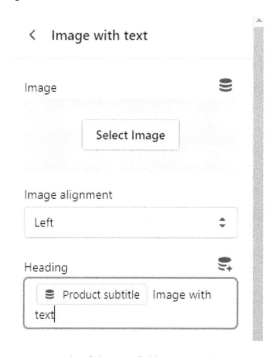

Figure 7.14 – Example of the metafields icon inside the theme editor

With that, we have learned how to dynamically update the values of the sections settings by using metafields, without having to create a new JSON template for each product. The only thing we have to do now is create the proper layout by arranging the necessary sections inside the JSON template.

Note that by using metafields, we can update all kinds of inputs, such as images, URLs, or even color swatches. However, we should create an appropriate metafield to ensure that the metafield value and the input-type value match.

As we saw previously, sections alone are pretty powerful. Mixing sections with JSON templates and metafields makes them even more impressive. It allows merchants to easily configure each page differently, without having to create new templates for each page or manually inserting sections.

Besides what we have learned so far, we know that Shopify also provides us with some additional section-specific tags that we can use to create even more powerful elements.

Exploring section-specific tags

While we can easily define styling or include the JavaScript code inside the theme's respective asset files, Shopify provides us with three types of tags that we can use to include CSS and JavaScript directly in the section file.

The stylesheet tag

Similar to the `schema` tag, the `stylesheet` tag is a Liquid tag that does not have any output on its own. It simply allows us to write CSS code inside the section files. Note that each section file can only contain a single `stylesheet` tag, which must stand on its own. It cannot be nested inside any other kind of Liquid tag:

```
{% stylesheet %}
{% endstylesheet %}
```

Although this might not look like a great idea at first, since we end up having CSS code spread over multiple files, Shopify will automatically collect all the CSS files from the different section files, combine them into one stylesheet file, and inject it into the theme file through the `content_for_header` global object.

> **Important note:**
> The `stylesheet` tag only accepts CSS values. We cannot include Liquid code inside the tag.

Note that bundled assets are not section- or block-specific. For section- and block-specific stylesheets, we will need to use the `style` tag.

The style tag

The `style` tag works similarly to the `stylesheet` tag as it allows us to write CSS code directly into the section file. However, the `style` tag has two significant differences compared to the `stylesheet` tag:

- The first difference is that the `style` tag is equal to using an HTML `<style>` tag, meaning that Shopify will not collect and bundle the CSS code that we included through the `style` tag. Instead, Shopify will render the HTML `style` tag and its content in the same place where we have included it:

```
{% style %}
{% endstyle %}
```

- The second and more important difference is that the `style` tag allows us to include Liquid code, which we can use to modify the CSS content by using the theme editor dynamically:

```
{% style %}
.featured-wrapper {
   background-color: {{ section.settings.background-
      color }};
}
{% for block in section.blocks %}
   {% case block.type %}
      {% when "product" %}
         .featured-product {

            font-size: {{ block.settings.product-font-size }}
               px;
         }
      {% when "collection" %}
         .featured-collection {
            font-size: {{ block.settings.collection-font-
                     size }}px;
         }
   {% endcase %}
{% endfor %}
{% endstyle %}
```

Note that as opposed to the input settings defined inside the `settings_schema.json` file, whose values can be accessed globally within any file, the section and block input values can only be accessed inside the section file itself or within the snippet file, after we pass the value as a parameter.

In the previous example, we saw how to output the section input values and use them to modify the CSS code dynamically. However, in our example, we have used a static class, meaning that the `background-color` CSS we applied will affect every single section and block similarly. *But what if we wanted to apply a different color to each section or block?*

To apply a unique styling to a specific element, we will need to use a `section` or `block` object, combined with the `id` attribute, to create a unique identifier that we can call later:

```
<div class="featured-wrapper featured-wrapper-{{ section.id
  }}"></div>
{% for block in section.blocks %}
  <div class="featured-collection featured-collection-{{
    block.id }}"></div>
{% endfor %}
```

Note that `section.id` will return a dynamically generated ID for dynamic sections and a section filename, without the Liquid extension for static sections. `block.id` will always return a dynamically generated ID.

Now that we have generated unique selectors, all we need to do is include the same selectors inside the `style` tag:

```
{% style %}
.featured-wrapper-{{ section.id }} {
  background-color: {{ section.settings.background-color}};
}
{% for block in section.blocks %}
  {% case block.type %}
    {% when "collection" %}
      .featured-collection-{{ block.id }} {

        font-size: {{ block.settings.collection-font-size
          }}px;
      }
  {% endcase %}
{% endfor %}
{% endstyle %}
```

Now that we know how to generate and call unique selectors, we can easily create different styling for both section and block elements. Note that it is also possible to include CSS with dynamic values in HTML using the `style` attribute.

The javascript tag

The javascript tag works in the same way as the stylesheet tag. The tag does not have any output on its own; it simply allows us to write JavaScript code directly inside the section files. We can only include a single javascript tag in a section, which must stand on its own, and it cannot be nested inside other Liquid tags:

```
{% javascript %}
{% endjavascript %}
```

Similarly, as with the stylesheet tag, the javascript tag only accepts JavaScript code. We cannot include any Liquid code inside the tag. Additionally, Shopify will automatically bundle any code inside the javascript tag and inject it into the theme file through the content_for_header global object.

Since we cannot use Liquid code inside the javascript tag, one way to apply section-specific JavaScript would be to use data attributes, which will output the specific input value and recover the value with JavaScript later:

```
<div class="rotating-announcement" data-speed="{{
  section.settings.speed }}"></div>
```

```
{% javascript %}
  var rotationSpeed = parseInt(document.querySelector
    ('.rotating-announcement').dataset.speed);
{% endjavascript %}
```

Considering that bundled assets are not section- or block-specific, if we needed to create some section- or block-specific JavaScript code, we would need to use an HTML <script> tag. Note that currently, Shopify does not have a Liquid tag that accepts Liquid code.

With the HTML <script> tag, we can now write section- and block-specific JavaScript and directly include the input settings values. However, note that when directly including the input settings values inside the JavaScript, you should include the values with whitespace control to ensure no extra whitespaces are included:

```
{{- section.settings.speed -}}
```

If you need to remind yourself how to use whitespace control, please revisit *Chapter 2, The Basic Flow of Liquid*, and consult the *Controlling whitespace* section, where we outlined the process of controlling whitespace.

Summary

In this chapter, we learned how to combine our previously attained knowledge of basic and specialized inputs with the section schema. This allows us to create everything from simple to complex layouts that we can easily configure throughout the theme editor.

By learning the difference between static and dynamic sections, we learned how to extend a section's functionality and make it accessible on any type of page. Additionally, besides learning how to create different types of sections, we have also gained knowledge of how to create different types of `blocks` modules, which will allow us to rearrange the layouts of a single section to make it easy to configure.

Lastly, we learned about the different types of section-specific tags that we can use to develop the content of the section further, as well as creating a unique experience for specific sections or even `blocks` modules.

In the next chapter, we will learn about what Shopify Ajax is and how we can utilize it to implement advanced functionalities and make a store more dynamic.

Questions

1. What is the main difference between the static and dynamic sections?
2. What object can we use to access the block input value? Write some code that will allow us to access the specific `blocks` module input value.
3. What is the difference between the `limit` and `max_blocks` attributes?
4. How can we apply section-specific CSS styling?

Practice makes perfect

As with the projects in the previous chapters, this project will contain detailed information about what we need to do and the appropriate instructions to help you achieve the results.

We recommend working on each project independently from the previous chapters since this will ensure you have truly understood what you have learned so far.

Not a single project has a correct or incorrect solution. However, if by any chance you get stuck, you can always consult the project solution, which can be found at the end of this book.

Project 4

For our fourth project, we will work on creating a section schema with multiple block types. While we can include any number of block types, we should make sure to include the following types:

- The text type, which will render a single rich text input
- The product type, which will render a single product element with a product name, price, image, and link
- The list type, which will render a one-level navigation list configurable from the Navigation admin section
- The video type, which will render a video from either the YouTube or Vimeo video platform

Here are the instructions for the assets:

1. Create a new section file called `featured-content.liquid`.
2. Create a separate snippet file for each section while also passing the proper object to each snippet.

The following are the instructions for this assignment:

1. Include all the necessary attributes that we need for a dynamic section.
2. Create the four block modules according to the specification provided.
3. Limit the video block type to a maximum of two repetitions, where all the other blocks should be limited to one occurrence.
4. The total number of blocks present at any time should not exceed four blocks.
5. We should configure each block's width separately by using the predefined drop-down values and apply the selected width to each block via a class, not by directly applying the value. The values are 100%, 50%, 33.33%, and 25%.
6. Only include the section or block-specific CSS inside the section file. We should include all other CSS stylings inside the theme's respective CSS file.
7. Define the following input settings inside the section block settings for each snippet file: `font-size`, `font-family`, text `color`, and `background-color`. We should be able to select the font family from the Shopify `font` library.
8. Create a text input element inside the section, which we will use to define the H1 heading for the entire section. If the input field is empty, we should hide the h1 HTML element.
9. Migrate the entire project into a `.json` page type template.

8
Exploring the Shopify Ajax API

In the previous chapters, we learned about the basics of Shopify and Liquid, which provided us with some solid groundwork for future development. After setting up a proper foundation for our future learning, we learned how Liquid core works. By learning about objects, tags, and filters, we have learned how to create complex functionalities using a somewhat simple and insignificant set of features. Lastly, we learned how to create easily configurable elements throughout the store using various input type settings, combined with the `sections` and `blocks` attributes.

Throughout these chapters, we have learned how to create elements with static content. *However, what if we were looking to update the content of our pages dynamically?* This is where the Shopify Ajax API comes to help. In this final chapter, we will go over the Shopify Ajax API, explain its requirements and limitations, as well as its possible use cases.

In this chapter, we will cover the following topics:

- Introduction to the Shopify Ajax API
- Updating the cart session with a POST request
- Retrieving data with a GET request

After completing this chapter, we will understand what the Shopify Ajax API is and the types of requests we can make, such as retrieving product information, adding products to the cart, or even reading the cart's current content. Additionally, we will learn about the typical uses cases for the Shopify Ajax API by working on some of our previous projects and improving them with the Ajax API.

Lastly, we will learn how to pull and render an automatically generated list of recommended products based on Shopify's algorithms, and then turn a general search input into a predictive search feature that's often requested by store owners.

Technical requirements

While we will explain each topic and present it with the accompanying graphics, we will need an internet connection to follow the steps outlined in this chapter, considering that **Shopify** is a hosted service.

The code for this chapter is available on GitHub at `https://github.com/PacktPublishing/Shopify-Theme-Customization-with-Liquid/tree/main/Chapter08`.

While this chapter will contain several real-life examples and use cases for each topic, we will need a basic understanding and knowledge of Ajax to be able to follow this chapter thoroughly.

Note that we will only show examples and work on projects related to the Shopify API and not Ajax in general. For detailed information on Ajax, we can consult `https://www.w3schools.com/js/js_ajax_intro.asp`, which provides a great introduction to Ajax.

The Code in Action video for the chapter can be found here: `https://bit.ly/2VUZ7Qp`

Introduction to the Shopify Ajax API

Ajax, or **Asynchronous JavaScript and XML**, is a method that we can use to exchange small amounts of data with the server and update the parts of any page, without the need to reload it in its entirety. *So, what exactly is the Shopify Ajax API?*

The Shopify Ajax API is a REST API endpoint through which we can send out requests to read or update certain information. For example, we can use a **GET** request to read the product or even the current cart data, or we can use a **POST** request to update the current content session of the cart.

Shopify Ajax is an unauthenticated API, which means that it does not require any tokens or API keys to gain access to store information. Shopify also provides us with an authenticated API named the **Shopify Admin API,** which apps and services use to communicate with Shopify servers.

Through the Shopify API, we can access most of our store data, whose responses will return JSON-formatted data, though we can't read customer and order data or update any store data – we can only do this using the Shopify Admin API.

Note that Shopify has certain rate limitations regarding the Ajax API to prevent abuse (sending an unlimited number of requests to Shopify servers). One such limitation is that there's a maximum input array size limit, currently limited to 250. Let's say that we are looking to pull information about all the products in a collection of over 1,000 products. We will have to use multiple queries to achieve this since we are limited to a maximum of 250 products per query.

> **Tip:**
> To keep everything concise and to the point, we won't mention all of the rate limitations here. For more information on Ajax API rate limitations, please refer to `https://shopify.dev/api/usage/rate-limits`.

Now that we have familiarized ourselves with the need-to-know Shopify Ajax API basics, we can learn more about the Ajax API from a practical standpoint.

Updating the cart session with a POST request

Previously, we mentioned that we can use a POST request to update the current cart session. Depending on the type of action we are looking to perform, we can pair the POST request with the following cart endpoints:

- `/cart/add.js`
- `/cart/update.js`
- `/cart/change.js`
- `/cart/clear.js`

While this might sound trivial, it is an essential aspect of today's e-commerce stores, where we expect to perform an action without refreshing an entire page.

The /cart/add.js endpoint

As its name suggests, the /cart/add.js endpoint allows us to add one or multiple product variants to the cart, without the need to refresh the cart. To perform this action, we need to create an array named items with an object inside containing the following two keys:

- The id key, whose value should contain the number type value of the variant ID we are adding to the cart.

- The quantity key, whose value should contain the number type value of the quantity we are adding to the cart.

If we need to include multiple variants, we can simply append multiple objects inside the items array:

```
items: [
    {
        id: 40065085407386
    },
    {
        id: 40065085603994,
        quantity: 5
    }
]
```

In the previous example, we can see an array with two objects containing a different set of variant id and quantity keys. However, note that the first object does not contain the quantity key. The reason for this is that the quantity key is entirely optional, and if we fail to include it, it assumes that the quantity value is equal to 1.

Let's look at how we could use this in a real-life example:

1. As you may recall, in *Chapter 4*, *Diving into Liquid Core with Objects*, and later in *Chapter 5*, *Diving into Liquid Core with Filters*, we worked on a Custom collection project by adding the additional collection to the collection template. However, the current functionality is that if we click on the **Add to Cart** button, the current form flow will add the respective variant to the cart and redirect us to the cart page. Let's change that by creating a POST request.

Before we include the Ajax API, we should review the current flow of the Custom collection form that we developed through *Chapter 4*, *Diving into Liquid Core with Objects*, and *Chapter 5*, *Diving into Liquid Core with Filters*. We can find the Custom collection form inside the Snippet directory, under collection-form.liquid:

```liquid
{% if product.compare_at_price != blank %}
<div class="custom-collection--item">
  <a href="{{ product.url }}">
    <img src="{{ product | img_url: "300x300" }}"/>
    <p class="h4 custom-collection--title">{{
        product.title }}</p>
    <p class="custom-collection--price">
      {{ product.price | money }}
      {% assign discount-price =
          product.compare_at_price |
              minus: product.price %}
      <span>Save {{ discount-price | money }}</span>
    </p>
    <span class="custom-collection--sale-badge">{{
      product.compare_at_price | minus: product.price |
        times: 100 | divided_by: product.compare_at_price
    }}%</span>
  </a>
  {% form "product", product %}
    <input type="hidden" name="id" value="{{
        product.first_available_variant.id }}" />
    <input type="submit" value="Add to Cart"/>
  {% endform %}
</div>
{% endif %}
```

2. As we can see, the collection form that we have created already contains the two necessary things we need: the `submit` button and the variant `id` that we have stored inside a hidden input. For more straightforward navigation, let's start by assigning a new class called `collection-submit` to the `submit` button:

```
{% form "product", product %}
  <input type="hidden" name="id" value="{{
    product.first_available_variant.id }}" />
  <input type="submit" class="collection-submit"
    value="Add to Cart"/>
{% endform %}
```

3. With the proper selector in place, we can now use an `addEventListener` on the `submit` button to capture the click event and pass the object to the function that we will create next:

```
const addSelector = document.querySelectorAll
  (".collection-submit");
if (addSelector.length) {
  for (let i = 0; i < addSelector.length; i++) {
    addSelector[i].addEventListener('click', function(e) {
      e.preventDefault();
      addCart(this);
    });
  }
}
```

4. In the preceding example, we created an `addSelector` constant for capturing the click event. Using `preventDefault()`, we canceled any current events' flows and passed the object of the clicked element to the `addCart` function. Now, let's look at creating the `addCart` function:

```
const addCart = (el) => {
  let formData = {
    'items': [
      {
        id: el.previousElementSibling.value
      }
    ]
```

```
    };
}
```

We started by creating an arrow function with an `el` parameter that we will pass the object of the previously clicked element to. Inside the `addCart` function, we created a local variable, inside which we assigned an array. This array contains an object that contains the `id` property of the variable we want to add to the cart.

5. Considering that we previously passed the clicked-on object to the arrow function, we used `previousElementSibling` to select the correct input element and return its value accordingly. Now that we have all the necessary assets in place, all we need to do is use the `fetch` request to POST the data to the Shopify server and update the current cart session:

```js
const addCart = (el) => {
  let formData = {
    'items': [
      {
        id: el.previousElementSibling.value
      }
    ]
  };

  fetch('/cart/add.js', {
    method: 'POST',
    headers: {
      'Content-Type': 'application/json'
    },
    body: JSON.stringify(formData)
  })
  .then(success => {
    console.log("Success:", success);
  })
  .catch((error) => {
    console.error('Error:', error);
  });
}
```

With that, we have successfully created a fully functional Ajax API POST request, allowing us to add any number of items to the current cart session without reloading the page. Additionally, we included the `then()` and `catch()` methods to return `success` and `error` messages inside the console log.

We also learned how to add specific products in selected quantities to the current cart session through the `/cart/add.js` endpoint. *However, what if we had certain line item properties on the specific product that we were looking to carry over to the cart?*

We can easily resolve this by simply including an additional parameter, `properties`, which accepts a key-value type object:

```
let formData = {
  'items': [
    {
      id: el.previousElementSibling.value,
      properties: {
        'Engraving message': 'Learning Liquid is fantastic!'
      }
    }
  ]
};
```

We should set the key so that it's equal to the name of the line item's input or the first part of the line item property, where the value should equal the value that was retrieved from the input or the second part of the line item property. Suppose we need to recall how line item properties work. In that case, we can revisit *Chapter 4, Diving into Liquid Core with Objects*, where, in the *Product customization* subtopic, located in the *Working with global objects* section, we explained how line item properties work.

If we need to pass a hidden line item that will only be visible in the order section part of the admin, we will need to append an *underscore* to the key:

```
let formData = {
  'items': [
    {
      id: el.previousElementSibling.value,
      properties: {
        '_Engraving message': 'Learning Liquid is fantastic!'
      }
    }
  ]
};
```

If we decide to use jQuery, we can make the code a lot more compact:

```
jQuery.post('/cart/add.js', {
  items: [
    {
      quantity: 1,
      id: 40065085407386,
      properties: {
        '_Engraving message': 'Learning Liquid is fantastic!'
      }
    }
  ]
});
```

However, we should check whether the theme we are working on already contains a jQuery library. Otherwise, we should avoid introducing a new library to the theme.

By covering both the JavaScript and jQuery solutions, we are now sure that we will be able to use our skills to produce the necessary Ajax API code. *However, what if we accidentally added a much higher quantity than we needed, and we need to update the product's quantity?*

The /cart/update.js endpoint

As its name suggests, the /cart/update.js endpoint allows us to update the line item values inside the current cart session.

While /cart/update.js works similarly to /cart/update.js, there are a few noticeable differences. For example, in /cart/add.js, we had to create a separate object when working with multiple variants, whereas with /cart/update.js, we only have to create a single object:

```
updates: {
    40065085407386: 5,
    40065085603994: 3
}
```

Notice that instead of using two sets of key values, we use only one here, where the key is represented by the variant ID and the quantity value represents the key value. Additionally, instead of items, we are now using updates. Let's create a function that will help us test out our new knowledge:

```
const updateCart = (el) => {
  let formData = {
    updates: {
      [el.previousElementSibling.value]: 5
    }
  };

  fetch('/cart/update.js', {
    method: 'POST',
    headers: {
      'Content-Type': 'application/json'
    },
    body: JSON.stringify(formData)
  })
  .then(success => {
    console.log("Success:", success);
  })
  .catch((error) => {
    console.error('Error:', error);
  });
}
```

As we can see, the general code for updating the cart session is similar to adding the product cart, which is not surprising. Besides updating the cart's content, `/cart/update.js` also allows us to add the product to the current cart session.

By using `/cart/update.js`, we can easily update the quantity of every item inside the cart by using the variant ID to identify which variant we are looking to update. *But what if the variant we are looking to update is not present inside the cart?* This is where the `/cart/update.js` alternate function triggers, updating the current cart session by adding the product variant to the cart with the selected quantity.

For example, in the previous `updateCart` function, we set the `quantity` value to a static value of 5. No matter how many times we call the preceding function, the total quantity of any variant inside the cart will never exceed 5. For this reason, we recommend always using `/cart/update.js` to update the existing cart items and `/cart/add.js` to add additional items to the cart.

With that, we have learned how to update the line items in the current cart session. However, as you may recall from *Chapter 4, Diving into Liquid Core with Objects*, in the *Product customization* subtopic, located in the *Working with global objects* section, we learned that it is possible to implement a different type of customization using line items. Consequently, this will sort the same product variants into different lines if their customization differs.

While these products might be on different lines, they will all have the same variant ID. *So, what will happen if we run* `/cart/update.js` *to update the specific variant on three different lines?*

The `/cart/update.js` endpoint would successfully perform its operation. However, since it does not know which line item we are looking to update, it will only update the first occurrence of the line item with the matching variant ID, and then it will stop. It will not update any additional occurrence with the same variant ID. *But what if we were looking to update a specific line item and not the first occurrence?*

The /cart/change.js endpoint

The `/cart/change.js` endpoint works similarly to the `/cart/update.js` endpoint as it allows us to update the line item inside the current cart session. However, two crucial differences are that we can only modify a single line item at a time and that (more importantly) we can specify exactly which line item we are looking to change.

Similar to the /cart/add.js endpoint, the /cart/change.js endpoint also uses an object with two key-value pairs – one to identify the line item and one to assign the needed quantity:

```
{
  'id': 40065085407386,
  'quantity': 7
}
```

While using id and the variant ID to identify the line item will not cause any errors, this will not resolve our problem as we can have multiple line items with the same variant ID in the cart. To resolve this, we can use the line property to identify the specific line item we want to change:

```
{
  'line': 3,
  'quantity': 7
}
```

The line value is based on the index position of the line items inside the current cart session, where the base value starts with 1. For example, if we have four items within the cart and we are looking to update the line item at the third position, we can set the line value to 3, as per our previous example. Note that the most common use for the /cart/change.js endpoint is to easily update the quantity of each line item inside the cart page.

Perform the following steps to implement the /cart/update.js endpoint successfully:

1. As we mentioned previously, to use the /cart/update.js endpoint successfully, we need two things: the quantity value, which we can quickly return from the input value that we modify, and the current position of the line item. To determine the position of the line item, we can use the JavaScript indexOf() method. Alternatively, we can introduce a data attribute and set its value to forloop. index if the quantity input is inside a Liquid for loop. We will use the second approach to add forloop.index as a data attribute here:

   ```
   <input type="number" name="quantity" value="0" data-
   quantityItem="{{ forloop.index }}"/>
   ```

2. After making sure that we have all the necessary attributes in place, all we need to do is use addEventListener to detect the change event on the input, and then pass the object to the changeCart() arrow function:

```
const changeCart = (el) => {
  let formData = {
    line: el.dataset.quantityItem,
    quantity: el.value
  };

  fetch('/cart/change.js', {
    method: 'POST',
    headers: {
      'Content-Type': 'application/json'
    },
    body: JSON.stringify(formData)
  })
  .then(success => {
    console.log("Success:", success);
  })
  .catch((error) => {
    console.error('Error:', error);
  });
}
```

The changeCart() arrow function is similar to the previous functions that we have created. The only difference is that now, we are using the /cart/change.js endpoint and no longer using static values for key-value pairs. Instead, we are pulling both values from the object that we passed previously.

While we can use both /cart/update.js and /cart/change.js to remove the item from the cart by simply setting the quantity value to 0, we would have to adjust the quantity of each line item to 0 manually. *But what if we wanted an easy way to clear out the entire cart with a single click?*

The /cart/clear.js endpoint

The `/cart/clear.js` endpoint is pretty simple to use compared to the previous endpoints as it does not accept any parameters. All we have to do is simply submit a POST request with `/cart/clear.js` and the cart will automatically clear all present items:

```
const clearCart = () => {
  fetch('/cart/clear.js', {
    method: 'POST',
    headers: {
      'Content-Type': 'application/json'
    }
  })
  .then(success => {
    console.log("Success:", success);
  })
  .catch((error) => {
    console.error('Error:', error);
  });
}
```

Note that if we were to run the preceding code on the cart, we would successfully clear all the items from the cart. However, we would still have to refresh the cart page to see the change because although we have cleared all the items from the current cart session, we have not removed the items from the actual DOM. We can implement a short `while` statement inside the `success` function and remove all line item elements to resolve this:

```
const cartItems = document.querySelector("[data-cart-line-
  items]");
while (cartItems.firstChild) {
  cartItems.removeChild(cartItems.firstChild);
}
```

Using the preceding code example, we have successfully removed all the items from our current cart session and the DOM. Notice that there are many additional fine-tuning aspects that we will need to handle, such as clearing the price, removing the cart table, and displaying a message stating that the cart is empty. However, to keep this book concise and to the point, we will not be getting into this, but you are free (and it is advisable) to keep upgrading the preceding code as you will only benefit from it.

So far, we have learned how to add products to the current cart session using the `/cart/add.js` endpoint, update the existing line items using `/cart/update.js` and `/cart/change.js`, and how to clear the current cart session using the `/cart/clear.js` endpoint. However, as we had a chance to see, while we could add easily, update, or even clear items from the current cart session, we still had to reload the page to see specific results, such as updating the item `counter` near the *cart* icon inside the header or the line item price when updating the item quantity.

While it would do so, we could quickly simply increment the item `counter` by the number of products we are adding to the cart. A more straightforward solution to achieve this is to use a GET request in combination with the Shopify Ajax API, which will allow us to retrieve all kinds of data from Shopify servers, including the number of products in the current cart session.

Retrieving data with a GET request

As we mentioned previously, using a GET request, we can pull all types of data from Shopify servers, except for customer and order information, which can only be accessed using an authenticated Shopify Admin API. Depending on the type of action we are looking to perform, we can pair the GET request with the following endpoints:

- `/cart.js`
- `/products/{product-handle}.js`
- `/recommendations/products.json`
- `/search/suggest.json`

The GET request is a pretty powerful method that we will commonly use in combination with a POST request to retrieve data after making changes to the current cart session. However, we can also use a GET request to retrieve and create complex functionalities, as we are about to learn.

The /cart.js endpoint

The /cart.js endpoint, as its name suggests, allows us to access the current cart session and retrieve all the information about the cart, as well as products inside the cart. We can use it to dynamically update the cart page or even create a cart drawer for the store and improve the purchase flow significantly. Let's take a look:

1. We can retrieve information about the cart using the following `fetch` method:

```
fetch('/cart.js')
  .then(response => response.json())
  .then(data => {
    console.log(data);
});
```

Note that the response of a successful GET request is a JSON object. The following example shows the response we will receive using the previous code to fetch the cart data:

```
{
    "token": "fd2d9bc86cfa72228b4de6bff52fe915",
    "note": null,
    "attributes": {},
    "original_total_price": 78594,
    "total_price": 78594,
    "total_discount": 0,
    "total_weight": 0,
    "item_count": 9,
    "items": [],
    "requires_shipping": true,
    "currency": "USD",
    "items_subtotal_price": 78594,
    "cart_level_discount_applications": []
}
```

As we can see, the JSON object provides us with all the necessary information about the cart and each product separately. Note that the variant information can be accessed through the `items` array object, which we have minified to keep everything concise.

2. Now that we have learned how to retrieve current cart session information, we can combine it with the POST request for /cart/add.js that we worked on previously, and then ensure that the cart counter is updated correctly each time we add a new product to the cart:

```
fetch('/cart/add.js', {
  method: 'POST',
  headers: {
    'Content-Type': 'application/json'
  },
  body: JSON.stringify(formData)
})
.then(success => {
  console.log("Success:", success);
  fetch('/cart.js')
  .then(response => response.json())
  .then(data => {
    document.querySelector("[data-cart-
        count]").innerHTML = data.item_count;
  });
})
.catch((error) => {
  console.error('Error:', error);
});
```

We now have all the knowledge necessary to retrieve different types of information from the current cart session. However, notice that the price values are pure strings within the cart response and have no currency format.

3. For example, let's say that we were looking to update the total price on the cart page every time we update the product quantity. To start, we will use the `fetch` method to retrieve the total price value:

```
fetch('/cart.js')
    .then(response => response.json())
    .then(data => {
        console.log(data.total_price);
});
```

While we successfully retrieved the price, all that we have received is an unformatted string value, which is not that useful to us:

```
78594
```

4. The easiest way to resolve this would be to look into how the theme developer has defined the currency formatting helper function throughout the theme. We can usually find it inside the theme `master js` file. In our case, this will be `theme.js`.

After identifying the keywords that we need, we simply need to apply the formatting to the value that we are looking to format:

```
fetch('/cart.js')
    .then(response => response.json())
    .then(data => {
        console.log(theme.Currency.formatMoney(
            cart.total_price, theme.moneyFormat));
});
```

Note that the formatting we have used in the previous example will work without any modifications in most cases. However, we might need to make some adjustments to specific themes – it all depends on how the theme developer defined the function.

Previously, we learned how to retrieve information about the current cart session and any data about any product inside the cart. *However, what if we wanted to retrieve product information more directly?*

The /products/{product-handle}.js endpoint

The /products/{product-handle}.js endpoint is a simple endpoint that we can use in combination with a GET request to retrieve information about any product in the store easily. Similarly, as with the /cart.js endpoint, /products/{product-handle}.js is relatively easy to use as it only requires us to include the handle of a product we are looking to retrieve data about:

```
const getProduct = (handle) => {
  fetch(`/products/${handle}.js`)
  .then(response => response.json())
  .then(product => {
    console.log(product.id);
  });
}
```

The return value that we will receive from the preceding example will include the product ID, which we will be using in the following example.

The most common use for this endpoint is when creating on-click functionalities, such as the quick view feature, where we need to load a lot of product information dynamically to avoid cluttering the DOM and slowing down the store.

The /recommendations/products.json endpoint

The /recommendations/products.json endpoint allows us to retrieve a list of JSON objects regarding the recommended products for the selected product based on Shopify algorithms, which we can use to construct a dynamic recommended section.

Through this endpoint, we can use three parameters, one of which is mandatory, while the other two are optional:

- The product_id parameter is a mandatory parameter whose value should be set to the ID of the product whose recommendation list we are looking to retrieve. Note that the product ID is not the same as the variant ID. They are two different attributes that we can retrieve through the product object.

- The limit parameter is an optional parameter that allows us to select the maximum number of recommended products we should receive per request. We cannot retrieve more than 10 recommended products per request due to Shopify limitations. This is the default value if we do not set the limit parameter.

- Last but not least is the `section_id` parameter, which, while optional, is quite an interesting parameter as it allows us to change the type of response that we will receive. By including the ID of a section as the `section_id` parameter value, we can select the parent element where we would like to render the recommended products. More importantly, we can also change the JSON response to an HTML string, which we can then use in combination with the `recommendations` object to output the recommended products dynamically.

Now that we have familiarized ourselves with all the attributes that we can use with the `/recommendations/products.json` endpoint, it is time to see them in action.

In the following example, we have used a `fetch` request, paired with the `/recommendations/products.json` endpoint, to generate a JSON object list and output them in the console log:

```
const productRecommendations = (productId, limit) => {
fetch(`/recommendations/products.json?product_id=
   ${productId}&limit=${limit}`)
   .then(response => response.json())
   .then(products => {
      console.log(products);
   });
}
```

As we can see, retrieving a JSON object for the recommended products is quite simple, as the only thing that we need to do now is pass the product ID to the `productId` parameter. As you may recall, the `limit` parameter is optional and will default to the maximum value of `10` when not included. Now, let's look at how we can include `section_id` and learn how to retrieve HTML strings instead.

Before we can modify the `fetch` request to accomplish this, we need to make specific preparations. The first thing we need to do is create a new section in the `Sections` directory. For our example, we will name it `recommended-products`.

Since we already have a recommended products section on the product page template and are only creating a new section to learn how this works, let's include this new section at the bottom of the `theme.liquid` layout file, just above the `</body>` tag. Now that we have created the section file and successfully included it, we must familiarize ourselves with the `recommendations` object.

As its name suggests, the recommendations object allows us to retrieve products from the product recommendations list. However, this particular object only works in combination with the /recommendations/products endpoint.

As we can see, the recommendations object is relatively simple to use as it only contains three attributes:

- The performed attribute returns a Boolean, depending on whether we have placed the recommendations object inside the section whose content we are rendering by combining the recommendations endpoint and the necessary parameters.

- The products_count attribute provides us with a number value for the number of products in the recommendation list.

- Last but not least, the products attribute allows us to retrieve an array of recommended product objects. We can combine the products attribute with the for tag to provide an output the same way as we did previously for the Custom collection project in *Chapter 5, Diving into Liquid Core with Filters.*

Let's return to the recommended-products section file we created and use the recommendations object to output the recommended products array:

```
<div class="product-recommendations">
  {% if recommendations.performed %}
    {% if recommendations.products_count > 0 %}
      {% for product in recommendations.products %}
        <div class="product">
          <a href="{{ product.url }}">
            <img class="product__img" src="{{
            product.featured_image | img_url: '300x300' }}"
            alt="{{ product.featured_image.alt }}" />
            <p class="product__title">{{ product.title }}</p>
            <p class="product__price">{{ product.price |
              money}}</p>
          </a>
        </div>
      {% endfor %}
    {% endif %}
  {% endif %}
</div>
```

With that, we have created a proper layout for the future recommendations list. However, if we were to preview the results on the theme, we would notice that the section does not render anything inside the product-recommendations div. As we mentioned previously, the recommendations object only works in combination with the recommendations endpoint, so let's look at how to use the endpoint to generate the necessary HTML strings to output the recommended product list.

To achieve this, we need to make a few adjustments to our previous fetch request:

1. The first thing that we need to do is include the additional parameters for the productRecommendations function that we will pass the section ID value to. Additionally, we will need to include the section_id parameter and its value to the fetch URL.

2. The second and more important step is to remove .json from the fetch URL. Otherwise, we will not be able to retrieve the JSON HTML string.

3. Last but not least, we will need to replace .json() with .text() inside the first then method.

At this point, we have all the necessary elements to retrieve the JSON HTML string. Let's test it out by calling the products inside the console log:

```
const productRecommendations = (productId, limit,
    sectionId) => {
fetch(`/recommendations/products?product_id=${productId}
    &limit=${limit}&section_id=${sectionId}`)
    .then(response => response.text())
    .then(products => {
        if (products.length > 0) {
            console.log(products);
        }
    });
}
```

However, before we can test this out, we need to pass the three values to our function:

- For `productId`, we can use the product ID value that we retrieved while learning about the `/products/{product-handle}.js` endpoint. Alternatively, we can use `product.id` inside any product template and copy the value that we receive.

- For `limit`, we can use any number value up to `10`, which is the maximum number of products we can receive as a response.

- For `sectionId`, we should include a string value equal to the name of the section we are looking to display the recommended products inside. In our case, the value is `recommended-products`.

The following is an example of passing all three values to our function:

```
productRecommendations(6796616663194, 3, "recommended-
   products");
```

If we were to preview our duplicate theme and check the console log inside the previous `fetch` function, we would see that we have successfully retrieved the JSON HTML string values for the recommended products.

Now that we have confirmed that everything works well, the only thing left to do is use the retrieved value and output the recommended products lists:

```
const productRecommendations = (productId, limit,
   sectionId) => {
fetch(`/recommendations/products?product_id=${productId}
   &limit=${limit}&section_id=${sectionId}`)
   .then(response => response.text())
   .then(products => {
     if (products.length > 0) {
       document.querySelector(".product-
           recommendations").innerHTML = products;
     }
   });
}
```

With that, we have successfully learned how to render a list of recommended products according to the layout that we previously defined inside the `recommended-products` section. Additionally, the product list will automatically update itself based on Shopify's algorithms.

While having a recommended list of products is a great feature for finding similar products, we still have to navigate to a specific product, and even then, we can't be sure that we will receive the exact results we needed. To help us with this, we can use a predictive search feature.

The /search/suggest.json endpoint

As its name suggests, the /search/suggest.json endpoint allows us to create a predictive search that will automatically provide us with a list of products that matches our query, either partially or completely.

Besides allowing us to use predictive search on products, we can also search for collections, pages, and even articles, depending on the type of parameters that we include. /search/suggest.json allows us to use seven different types of parameters. However, to keep everything concise and to the point, we will only cover the most important ones that are needed to make the predictive search functionality work:

- The first parameter on our list is the q parameter, which is a mandatory string type parameter whose value should be equal to the search query.

- The type parameter allows us to specify the type of result we are looking to receive. We can include the following comma-separated values: product, page, article, and collection. The type parameter is also mandatory.

- The limit parameter is an optional integer parameter that allows us to set the number of results we should receive per request. Note that if we do not include the limit attribute, it defaults to 10, which is the maximum number of results we can receive per request.

- The resources attribute is a mandatory hash type parameter that requests resource results for the query based on the type and limit fields.

In the following example, we have used a fetch request, paired with the /search/suggest.json endpoint, to generate a JSON object list that matches our search query and outputs it in the console log:

```
const predictiveSearch = (query, limit, type) => {
fetch(`/search/suggest.json?q=${query}&resources[type]=
  ${type}&resources[limit]=${limit}`)
  .then(response => response.json())
  .then(suggestions => {
    const productSuggestions =
        suggestions.resources.results.products;
```

```
    if (productSuggestions.length > 0) {
        console.log(productSuggestions);
    }
    });
}
```

As we can see, retrieving predictive search results based on the search query is relatively simple, since the only thing we need to do now is pass down the required values to our function:

```
const searchSelector = document.querySelectorAll
    (".search-bar__input");
if (searchSelector.length) {
    for (let i = 0; i < searchSelector.length; i++) {
        searchSelector[i].addEventListener('input',
            function(e){
            e.preventDefault();
            predictiveSearch(this.value, 4,
                "product,page,article,collection");
        });
    }
}
```

By typing a search query inside the search field, we will notice that we have successfully retrieved a combined list of up to four products, pages, articles, or collections of JSON objects that partially or fully match our search query inside the console log. The only thing left to do now is use the response values to generate the results inside the DOM.

With the `/products/{product-handle}.js` endpoint, we had a parameter that allowed us to retrieve a JSON HTML string to output the results into the DOM easily. This is not the case with the `/search/suggest.json` endpoint, however; to render these results, we will need to use JavaScript to create the layout and functionality that we need. To keep everything concise and to the point, we will not be covering that in this book. However, we recommend finishing the project as it will be some excellent practice that will help you with everything you have learned so far.

For additional information on predictive search parameters and their general requirements and limitations, please consult `https://shopify.dev/api/ajax/reference/predictive-search`.

Summary

Throughout this final chapter, we have familiarized ourselves with the Shopify Ajax API and learned about different types of use cases. First, we learned how to upgrade the current purchase flow using the `/cart/add.js` endpoint, through which we can add any number of products, quantities, and line item customizations, whether they are public or private, directly to the current cart session.

By learning how to handle the `/cart/change.js` endpoint, we gained the necessary knowledge to create a feature that includes a specific product and quantity, such as an automatic gift or upsells feature. Using `/cart/update.js` in combination with the `/cart.js` endpoint, we learned how to update the cart's content dynamically and retrieve it. We can then use this to create a cart drawer feature.

Additionally, we learned how to use the `/products/{product-handle}.js` endpoint to retrieve an automatic list of recommended products and render their content into a section of our choice.

Lastly, we learned about the `/search/suggest.json` endpoint, which allows us to create a predictive search functionality, one of the most requested features by store owners.

From the very beginning of this book, we have worked together on pushing the limits of our knowledge and creating a solid flow of understanding that will help us on our path of becoming a Shopify expert. While we haven't gone over every piece of Liquid code, we have worked on some exciting projects where we have learned about something a lot more beneficial. Our goal was not to simply create a list of where we would list all the different methods and attributes, which we can always find by looking through the Shopify documentation, but also to learn how both Shopify and Liquid work.

While it suffices to say that through the knowledge we've gained here, we should be ready to start working on the Shopify theme independently, note that our adventure is not ending – it is only just beginning.

Shopify is a constantly evolving platform, and it will require us to stay up to date with all the latest announcements and approaches. Luckily, Shopify offers various communities to improve our knowledge further or get assistance from other Shopify experts on various topics. Last but not least, we have a Discord channel at our disposal, where we can talk with other developers and both get assistance when we need it or share our knowledge with other developers: `https://discord.gg/shopifydevs`.

Further reading

- Shopify official documentation: `https://shopify.dev/`

- Shopify cheat sheet: `http://cheat.markdunkley.com/`

- Developer changelog: `https://shopify.dev/changelog`

- Official community: `https://community.shopify.com/`

- Twitter announcements: `https://twitter.com/shopifydevs`

- Shopify Developer YouTube channel: `https://www.youtube.com/channel/UCcYsEEKJtpxoO9T-keJZrEw`

- Shopify official blog for all the latest information about the world of Shopify: `https://www.shopify.com/partners/blog`

Assessments

Chapter 1, Getting Started with Shopify

Question 1

What is the Partners program?

Answer

The Partners program is a platform created by Shopify that assembles people from all over the world. Through this platform, we can build new ecommerce stores for store owners, design themes, develop apps, refer new clients to Shopify, and most importantly, create a development store for us to practice. We can remind ourselves of this within the *How to start?* topic.

Question 2

How can we disable the password protection of the development type store?

Answer

We can disable our password protection store by clicking on the **See store password** button within the banner on the **Themes** section, located under the Online store section, or by selecting the **Online store** and subsequently clicking the **Preferences** inside the expanded dropdown. Once inside, we can easily remove the password protection on the regular store. However, since our store is in development mode, this option is currently disabled. We can remind ourselves of this within the *Sidebar* subtopic, in the *Understanding theme structure* topic.

Question 3

What is the difference between the Layout and Templates directory files?

Answer

The **Layout** directory is the main directory of our theme. It contains the essential files, and it is where all other files, including the template files, will render. The template files are a group of files that allow us to easily create and manage the look of multiple pages all at once. We can remind ourselves of this within the *Sidebar* subtopic, in the *Understanding theme structure* topic.

Question 4

Under what circumstances will the new template file be visible inside the admin section of your page?

Answer

Considering that the admin side of Shopify can only read values from the currently published theme, we must meet two conditions. Besides creating a new template file, we also need to ensure that the same template file exists within the currently live theme or to publish our duplicate theme live. We can remind ourselves of this within the *Templates* subtopic, in the *Understanding theme structure* topic.

Question 5

What type of files and under what conditions will allow us to access the variables within the parent file scope?

Answer

The **Snippets** files allow us to re-use repetitive pieces of code over **Templates/Sections** by referencing their names. Besides allowing us to re-use parts of code, the **Snippets** will enable us to access the variables inside the parent element for as long as we pass those variables to the snippet as parameters.

Chapter 2, The Basic Flow of Liquid

Question 1

What type of delimiter should we use if we are expecting an output as a result?

Answer

If we expect output from Liquid code, we should use a double bracket delimiter, as we should only use a bracket with a percentage when performing a certain logic. We can remind ourselves about this within the *Understanding Liquid and its delimiters* section.

Question 2

What will the result of the following conditional be, and why?

```
{% if collection.all_products_count > "20" %}
   The number of products in a collection is greater
     than 20!
{% endif %}
```

Answer

Considering that `collection.all_products_count` by default returns a number as its value, where the value we are comparing it against is a string since it is encapsulated inside the parentheses. Since we cannot compare values of different types, the conditional will return `false`, and our message will not be shown. We can remind ourselves about this within the *Learning the comparison operators* section.

Question 3

What are the two methods to access an item inside an array?

Answer

We can access the items inside an array using two methods. The first method allows us to use the index position of an item to recover the exact item that we are looking for, while the second method allows us to loop over all of the items inside an array. We can remind ourselves about this within the *Array* subsection, in the *Understand the type of data* section.

Question 4

What is the correct way of accessing an object using its handle?

Answer

We can access the object using its handle by pluralizing the object's name we are trying to access, followed by either a squared bracket ([]) or dot (.) notation. Both methods of accessing the object are correct. However, they each have their use. We can remind ourselves about this within the *EmptyDrop* subsection, in the *Understand the type of data* section.

Question 5

What are the two problems inside the following block of code?

```
{% if customer != nil %}
   Welcome {{- customer.name -}} !
{% endif %}
```

Answer

Since nil is a special data type that returns an empty value, it does not have a visual representation, which is the first problem. We can remind ourselves about this within the *Nil* subsection, in the *Understand the type of data* section. The second problem is that we have added a hyphen to both sides of our customer.name output. While the hyphen on the right side of our output will clear the unwanted whitespace before the exclamation mark, we have also added a hyphen on the left side, removing the spacing between the word "Welcome" and our customer's name. We can remind ourselves about this within the *Controlling the Whitespace* section.

Chapter 3, Diving into Liquid Core with Tags

Question 1

What parameters should we use inside a for loop if we want to show a maximum of seven iterations while also skipping the first three iterations?

Answer

If we are looking to create a loop that will skip the first three iterations and output a maximum of seven iterations, we should use a combination of the offset and limit parameters. The offset tag will allow us to skip any number of iterations, depending on the value we assign to it. The limit parameter will allow us to limit the number of iterations the tag should perform. We can remind ourselves of this by revisiting the *for parameters* subsection of the *Iterations tags* section.

Question 2

What types of data can we assign to a variable created using the capture tag?

Answer

We can assign any type of data to a variable created using the capture tag. However, a variable created using the capture tag will always return string data as a result. We can remind ourselves of this by revisiting the *capture* subsection of the *Variable tags* section.

Question 3

What are the two problems in the following block of code?

```
liquid for product in collections["outdoor"].products
  if product.price > 10000
    continue
  else
    product.title
  endif
endfor
```

Answer

While using the liquid tag allows us to eliminate the brace delimiters within the code block we should not remove them from the liquid tag. We should place an opening brace delimiter with a percentage symbol on the left side of the liquid tag. The second issue with our code is that we are missing an echo tag in front of the product.title, which replaces the double curly brace delimiters. We can remind ourselves of this by revisiting the *The liquid and echo tags* subsection, in the *Theme tags* section.

Question 4

What approach should we take to modify an HTML-generated product form by replacing the existing class attribute with a combination of a string and a variable?

Answer

Considering that the `form` tag does not accept a combination of strings and variables as its parameter, we should first assign these values to a variable using the `capture` tag and then pass it to the `form` tag. We can remind ourselves of this by revisiting the *form* subsection, under the *Theme tags* section.

Question 5

What parameter should we use if we want to pass an object from the parent element?

Answer

The only tag that allows us to pass down objects from the parent element is the `render` tag, and even then, we can only do so by using the `with` and `as` parameters. We should set the value of the `with` parameter to the object we are looking to pass, and the value of the `as` parameter should be the name of the variable we will use within our snippet file. We can remind ourselves of this by revisiting the *render* subsection, in the *Theme tags* section.

Chapter 4, Diving into Liquid Core with Objects

Question 1

What are we missing in the following block of code to make `form` functional?

```
{% form "product", product %}
  <input type="hidden" value="{{
   product.first_available_variant.id }}" />
    <input type="submit" value="Add to Cart"/>
{% endform %}
```

Answer

While we have introduced the `id` variant, which is necessary to create a working product form, we didn't use a `name` attribute with `id` as its value. We can remind ourselves of how this works by going back to the *Custom collection* subtopic, in the *Working with global objects* section.

Question 2

How can we get access to the `product` object through a link defined in the admin navigation?

Answer

To access the `product` object through the navigation menu, we will need to use a `for` tag to iterate over the navigation menu. Once we have found which menu item is `product_type`, we can use that menu item, followed by the `object` attribute, to access that specific `product` object. We can remind ourselves of how this works by going back to the *Custom navigation* subtopic, in the *Working with global objects* section.

Question 3

What are the two approaches to accessing single and multiple `metafield` objects?

Answer

We can access the single `metafield` object using the object of the page we are looking to recover `metafield` from, followed by the `metafields` object, followed by `namespace`, and finally followed by `key`. If we are looking to recover multiple `metafields` objects, we will need to use a `for` tag to iterate over all the metafields with their namespace. We can remind ourselves of how this works by going back to the *Improving the workflow with metafields* section.

Question 4

What adjustment do we need to make to the input element if we were looking to capture the line_item value and hide it on the checkout page?

```
<input type="text" name="properties[Your Name]"
placeholder="Your Name"/>
```

Answer

If we were looking to capture the line_item value, we would need to introduce an underscore as a first character inside the square bracket. By introducing this underscore within the line_item input, we will automatically hide the specific line_item from the checkout page. However, this will not hide it on the cart page. The cart page will require some manual adjustments. We can remind ourselves of how this works by going back to the *Custom navigation* subtopic, in the *Working with global objects* section.

Chapter 5, Diving into Liquid Core with Filters

Question 1

Suppose that we have an array named product_handles with handles of 30 products. What issue in the following code would prevent us from outputting the images of all 30 products successfully?

```
{% for handle in product_handles %}
  {% assign product_object = all_products[handle] %}
  {% for image_item in product_object.images %}
    <img src="{{ image_item | img_url }}"/>
  {% endfor %}
{% endfor %}
```

Answer

Since we are looking to output more than 20 products, in this case, 30 products, we cannot use the all_products object, as the all_products object has a limitation that we can only call it 20 times on a single page. If we are looking to recover data from more than 20 products, we need to assign them to a collection and then perform a loop over those products. We can remind ourselves of the all_products object by visiting the *Working with HTML and URL filters* section.

Question 2

Why is only using the `model_viewer_tag` tag not recommended when creating the product media gallery?

```
{% for media in product_object.media %}
  {% case media.media_type %}
    {{ media | model_viewer_tag }}
  {% endcase %}
{% endfor %}
```

Answer

While `model_viewer_tag` will correctly output the necessary HTML media tag for each media type, we should only use `model_viewer_tag` as a fallback if all other media tags fail to render the proper tag. Using `model_viewer_tag` will prevent us from including any of the specific parameters for each `media` tag. We can remind ourselves of the `media` object by looking over one of the previous projects we have completed in this chapter, enhancing the product media gallery.

Question 3

Which filter could we use if we were looking to access an item at a specific location inside the array?

Answer

If we are looking to access an item at a specific location, we will need to use an array type filter named `index`. Using the `index` filter, we can access the specified index location in an array and return its value. We can remind ourselves of the index parameter by looking over one of the previous projects we have completed in this chapter, product accordions.

Question 4

What filter can we use to quickly update any occurrence of a string value inside the theme files?

Answer

To easily update any occurrence of a string value, we will need to use the t (translation) filter. By defining the translation keys, we can quickly update or even translate any string value without the need to update the hardcoded string values across multiple files manually.

Chapter 6, Configuring the Theme Settings

Question 1

What are the two types of input settings?

Answer

The first set of settings is called basic input types, and it consists of six types of settings, which allow us to output basic HTML input elements through which we can dynamically output certain content. The second set of settings, otherwise called specialized settings, allows us to generate specialized selector type fields to access various objects through the store and output their content using their attributes. We can remind ourselves of the basic and specialized input types of settings by visiting the *Basic input types* and *Specialized input settings* sections in *Chapter 6, Configuring the Theme Settings*.

Question 2

What's the issue that will cause an error with the following piece of code?

```
{
  "type": "text",
  "id": "header_announcement",
  "label": "Text",
}
```

Answer

While the code structure is correct, we have accidentally included an extra comma after the last attribute inside the text setting type, which will cause a JSON error and prevent us from saving the changes. We can remind ourselves of the strict JSON format by visiting the *Basic JSON settings* section in *Chapter 6, Configuring the Theme Settings*.

Question 3

How can we include a custom font file within Shopify and use it throughout the theme editor?

Answer

While `font_picker` allows us to access a significant number of fonts within Shopify, we have no way of including the custom font in this library. To include a custom set of fonts, we will have to use a `select` input type of settings, where we can manually create a list of fonts we wish to include. We can remind ourselves of how to include custom fonts by visiting *The select input* subsection in the *Basic input types* section from *Chapter 6, Configuring the Theme Settings*.

Question 4

What are the two issues that will prevent us from executing the following piece of code?

```
{
    "type": "range",
    "id": "number_of_products",
    "min": 110,
    "max": 220,
    "step": 1,
    "unit": "pro",
    "label": "Number of products",
    "default": 235
}
```

Answer

Considering that each `range` slider can have a maximum of 100 steps, the first issue is that the `min` and `max` attribute values are too far apart, which we can resolve by decreasing one of the two values so that they do not have more than 100 steps. An alternative solution is to increase `step` to a higher value, consequently reducing the number of steps between the `min` and `max` values.

The second issue is that the `default` attribute value is currently exceeding the `max` attribute value. We can resolve this by decreasing the `default` value or increasing the `max` attribute value. After ensuring that the `default` attribute value does not exceed the `max` value, we also need to ensure that the `default` value is also higher than the `min` attribute value. We can remind ourselves of the `range` type setting format by visiting *The range input* subsection in the *Basic input types* section from *Chapter 6, Configuring the Theme Settings*.

Chapter 7, Working with Static and Dynamic Sections

Question 1

What is the main difference between the static and dynamic sections?

Answer

The main difference between the static and dynamic sections is that we can only add the dynamic section to the JSON templates and the home page using the **Add section** button inside the theme editor. Additionally, we can repeat this any number of times with different content.

On the other hand, the static section needs to be included manually inside a theme template using the `section` tag. We can include the same static section inside multiple templates. However, each section will display the same content as we can only have one instance of a static section. We can remind ourselves of static and dynamic sections by visiting the *Static versus dynamic sections* section.

Question 2

What object can we use to access the block input value? Write some code that will allow us to access the specific `blocks` module input value.

Answer

We can access the block's input value through the `section` object and combine it with the `blocks` attribute, which will return an array of block objects:

```
{% for block in section.blocks %}
  {% case block.type %}
    {% when "block-type" %}
    {{ block.settings.input-id }}
  {% endcase %}
{% endfor %}
```

We can remind ourselves how to access the block input types by visiting the *Building with blocks* section.

Question 3

What is the difference between the `limit` and `max_blocks` attributes?

Answer

The main difference between the `limit` and `max_blocks` attributes is that the `limit` attribute only allows us to limit how many times we can repeat a particular block type. The `max_blocks` attribute, on the other hand, allows us to limit how many blocks we can include inside a particular section, regardless of the block type. We can remind ourselves how to use the `limit` and `max_blocks` attributes and the differences between the two by visiting the *Building with blocks* and *The max_blocks attribute* sections.

Question 4

How can we apply section-specific CSS styling?

Answer

If we need to include a section-specific CSS, we can do this using the `{% style %}{% endstyle %}` tag, which will allow us to use Liquid code. However, besides the `style` tag, we will also need to define a unique identifier that we will call later as the selector. We can do this using the `section` object and `id` attribute, which will return a dynamic ID for the dynamic section or a section filename for the static section. We can remind ourselves how to create section-specific CSS by visiting *The style tag* section.

Please note that the solutions for all the four projects are available on GitHub: `https://github.com/PacktPublishing/Shopify-Theme-Customization-with-Liquid/tree/main/Projects`

Appendix

Frequently Asked Questions

Through the years of helping developers adapt to Liquid, specific situations and questions have arisen to which answers are not so easy to come by. We will now list the most frequent situations and offer some advice that will help improve your knowledge of both Shopify and Liquid and make your work that much easier:

1. *Is it possible to find the location of some particular code inside the code editor without manually searching every template, section, or snippet?*

   One of the biggest problems that the developers have when starting to work with **Shopify** is that they have trouble finding a specific piece of code across different templates, sections, and snippets. While after some time, we do get used to finding things more efficiently, it can be pretty troublesome at the start.

   To help us with this, we can use a nifty little **Chrome** addon called **Shopify Theme Search by Bold**. This little addon provides us with a search input inside the code editor to type in any string that we are looking for. After a few seconds, it will highlight every directory and file that contains any occurrence of the searched string. You can read more about it at `https://chrome.google.com/webstore/detail/shopify-theme-search-by-b/epbnmkionkpliaiogpemfkclmcnbdfle`.

2. *No matter how many new duplicate themes we create, for some reason, the duplicate theme does not have any of the previous customizations saved. What should we do in this case?*

    The second most problematic issue happens during the theme duplication process. After creating a theme duplicate, we may notice that the duplicate theme has reset entirely. It does not contain any of the customizations that we had previously set inside the theme editor.

    The problem lies inside the `settings_data.json` file, which, if we open it, we will see is empty, and in some instances, the file might be missing altogether. The reason for this occurrence is apparent – Shopify did not copy the file's content correctly – but the question is *why*? If we were to navigate to the theme from which we initially created a duplicate, open the `settings_data.json` file, and try to copy its content to the new duplicate theme, we would not be able to save the file due to an error inside the JSON file. If we try to save the copied content to a new file, Shopify will not allow it and instead provide us with an error code of the problematic value.

    To resolve this, we will need to manually search for every value that we received inside the error and update it accordingly. Possible issues could be that we have entirely deleted a particular section block without first removing the block from inside the theme editor. Shopify will keep trying to load the block. However, since we have deleted it, it will not be able to, and it will break the JSON file. The second most frequent issue comes up when using the `range` input type, where the current value of the input exceeds the range between the min and max values. Only after resolving all of these issues will we be able to save the file correctly.

3. *Is it possible to recover a deleted theme?*

    Each store allows us to keep up to 20 duplicate themes as a backup and up to 50 duplicate themes on the **Shopify Plus** plan. If we delete a theme file for any reason and do not have a backup file saved on GitHub or somewhere locally, the file will be permanently lost, and we will not be able to recover it in any way.

4. *How can we find which app is being used by a particular store?*

    While it is not possible to see every type of app that a store is using, we can retrieve the names of most of the apps by using the **Shopify App Detector by Fera.ai** or **Koala Inspector - Inspect Shopify Shops** Chrome addons. You can find more info on Shopify App Detector by Fera.ai using the following link: `https://chrome.google.com/webstore/detail/shopify-app-detector-by-f/lhfdhjladfcmghahdbcmlceajdlbkale`.

You can find more info on Koala Inspector - Inspect Shopify Shops using the following link: `https://chrome.google.com/webstore/detail/koala-inspector-inspect-s/hjbfbllnfhppnhjdhhbmjabikmkfekgf`.

5. *How can we easily identify the Liquid code that is slowing down the store?*

   To help us with this, Shopify has introduced an addon named **Shopify Theme Inspector for Chrome**, which we can use to retrieve a detailed analysis render profiling and help us narrow down which files are causing the most significant delay. We can find more info on Shopify Theme Inspector for Chrome using the following link: `https://chrome.google.com/webstore/detail/shopify-theme-inspector-f/fndnankcflemoafdeboboehphmiijkgp`.

6. *When we try to send a collaborator account request to the client using the Shopify partners platform, for some reason, we keep getting the message that the store URL is not associated to a shop. How can we resolve this?*

   The reason for this is that in most cases, store owners have set up a custom domain whose URL is quite different than the `myshopify.com` URL that we require to send them a collaborator account request, so replacing the `.com` with `.myshopify.com` will not resolve our problem. We can easily resolve this by simply navigating to the URL they have provided us by opening the inspector and typing `Shopify` inside the console, which will prompt a set of various information about the store. Among that, we will find a correct `myshopify.com` URL. Note that the `Shopify` keyword is case-sensitive.

7. *Does the Shopify code editor have a dark mode?*

   The answer is yes! But it is pretty hidden. In order to reveal it, navigate to the code editor and click on the button to expand the editor screen. We can find it in the top-right corner of the editor screen. Once we have expanded the editor screen, we will notice that we now have two scroll bars on our right. If we scroll down to the bottom of the page, we will find the **White** and **Black** buttons allowing us to change the color mode of the editor. However, note that the editor will revert to its default color when we close the expanded editor view.

8. *Is there a way to automatically format the code inside the Liquid files?*

   The answer is yes! We can highlight the entire Liquid file inside the code editor using the *CTRL + A* and then press *Shift + Tab*, automatically formatting the entire file accordingly. However, while auto-formatting will be helpful with most files, note that automatic formatting does not work on the `section` schema tag and its content.

Packt.com

Subscribe to our online digital library for full access to over 7,000 books and videos, as well as industry leading tools to help you plan your personal development and advance your career. For more information, please visit our website.

Why subscribe?

- Spend less time learning and more time coding with practical eBooks and Videos from over 4,000 industry professionals

- Improve your learning with Skill Plans built especially for you

- Get a free eBook or video every month

- Fully searchable for easy access to vital information

- Copy and paste, print, and bookmark content

Did you know that Packt offers eBook versions of every book published, with PDF and ePub files available? You can upgrade to the eBook version at packt.com and as a print book customer, you are entitled to a discount on the eBook copy. Get in touch with us at customercare@packtpub.com for more details.

At www.packt.com, you can also read a collection of free technical articles, sign up for a range of free newsletters, and receive exclusive discounts and offers on Packt books and eBooks.

Other Books You May Enjoy

If you enjoyed this book, you may be interested in these other books by Packt:

Responsive Web Design with HTML5 and CSS - Third Edition

Ben Frain

ISBN: 978-1-83921-156-0

- Integrate CSS media queries into your designs; apply different styles to different devices
- Load different sets of images depending upon screen size or resolution
- Leverage the speed, semantics, and clean markup of accessible HTML patterns
- Implement SVGs into your designs to provide resolution-independent images
- Apply the latest features of CSS like custom properties, variable fonts, and CSS Grid
- Add validation and interface elements like date and color pickers to HTML forms
- Understand the multitude of ways to enhance interface elements with filters, shadows, animations, and more

Modernizing Enterprise CMS Using Pimcore

Daniele Fontani , Marco Guiducci , Francesco Minà

ISBN: 978-1-80107-540-4

- Create, edit, and manage Pimcore documents for your web pages
- Manage web assets in Pimcore using the digital asset management (DAM) feature
- Discover how to create layouts, templates, and custom widgets for your web pages
- Administer third-party add-ons for your Pimcore site using the admin UI
- Discover practices to use Pimcore as a product information management (PIM) system
- Explore Pimcore's master data management (MDM) for enterprise CMS development
- Build reusable website components and save time using effective tips and tricks

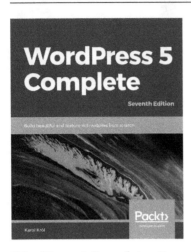

WordPress 5 Complete - Seventh Edition

Karol Król

ISBN: 978-1-78953-201-2

- Learn to adapt your plugin with the Gutenberg editor
- Create content that is optimized for publication on the web
- Craft great looking pages and posts with the use of block editor
- Structure your web pages in an accessible and clear way
- Install and work with plugins and themes
- Customize the design of your website
- Upload multimedia content, such as images, audio, and video easily and effectively
- Develop your own WordPress plugins and themes
- Use WordPress to build websites that serve purposes other than blogs

Packt is searching for authors like you

If you're interested in becoming an author for Packt, please visit `authors.` `packtpub.com` and apply today. We have worked with thousands of developers and tech professionals, just like you, to help them share their insight with the global tech community. You can make a general application, apply for a specific hot topic that we are recruiting an author for, or submit your own idea.

Share Your Thoughts

Now you've finished *Shopify Theme Customization with Liquid*, we'd love to hear your thoughts! Scan the QR code below to go straight to the Amazon review page for this book and share your feedback or leave a review on the site that you purchased it from.

https://packt.link/r/1-801-81396-5

Your review is important to us and the tech community and will help us make sure we're delivering excellent quality content.

Index

Symbols

/cart/add.js endpoint 266-271
/cart/change.js endpoint 273-275
/cart/clear.js endpoint 276, 277
/cart.js endpoint 278-280
/cart/update.js endpoint 272, 273
/products/{product-handle}.
    js endpoint 281
/recommendations/products.
    json endpoint 281-283
/recommendations/products.
    json endpoint, parameters
  limit 281
  product_id 281
  section_id 282
/search/suggest.json endpoint 286, 287

A

additional filters
  about 170
  exploring 170
  reference link 172
admin panel
  core aspects 9
  navigating 8
  sales channel 10-12
  settings 12
Ajax API rate limitations
  reference link 265
and tags 62, 63
array 42
article input 211, 212
assign tag 66, 67
Asynchronous JavaScript and
    XML (Ajax) 264

B

basic input types
  about 184
  checkbox input 185, 186
  number input 186
  radio input 187, 188
  range input 188, 189
  select input 190, 191
  textarea input 193
  text input 192
block object
  reference link 241
blocks
  building with 234-242
  max_blocks attribute 242, 243

blog input 212, 213
Boolean 39
break statement 74

C

capture tag 67, 68
cart session
  updating, with POST request 265
case tags 63, 64
checkbox input 185, 186
checkout.liquid
  reference link 26
class attribute 230
code editor
  Header section 20
  Sidebar section 22
collection input 213, 214
collection page
  creating 57, 58
color input 199, 200
comment tag 91
comparison operators
  <= operator 36
  < operator 35
  == operator 34
  != operator 34
  >= operator 36
  > operator 34
  about 34
config directory 28
Content Delivery Network (CDN) 99, 132
content_for_header object 125
content_for_index object 125
content_for_layout object 125
content objects
  about 124
  content_for_header object 124

content_for_index object 125
content_for_layout object 125
continue statement 73
control flow tags
  about 58
  and tags 62, 63
  case tags 63, 64
  if/else/elsif tags 58-62
  or tags 62, 63
  unless tags 65
  when tags 63, 64
customer privacy
  reference link 12
cycle group 80
cycle tag 78-81

D

data
  retrieving, with GET request 277
data types
  about 38
  array 42
  Boolean 39
  EmptyDrop 43, 47
  nil 40, 41
  number 39
  string 38
decrement tag 70, 71
default filter 170
deprecated settings
  about 219
  font input 219
  snippet input 220
deprecated tags 92
dynamic content 32
dynamic sections 227, 228

E

echo tag 83, 84
else tags 58-73
elsif tags 58-62
EmptyDrop data type 43, 47

F

Facebook Pixel
  reference link 12
filter 100
float 39
font filters
  reference link 210
font input 219
font objects
  reference link 210
font_picker input 206-210
footer element 82
form tag 85-87
for parameters
  about 74
  limit parameter 75
  offset parameter 76
  range parameter 77
  reversed parameter 78
for tags 71-73

G

GET request
  /cart.js endpoint 278-280
  /products/{product-handle}.
    js endpoint 281
  /recommendations/products.
    json endpoint 281-286

/search/suggest.json endpoint 286, 287
  used, for retrieving data 277
global objects
  custom collection 98-104
  custom navigation 105-111
  product customization 111-117
  working with 96, 97
Google Analytics
  reference link 12

H

handle 43-46
header element 82
Header section
  about 20
  customize theme button 21
  expert theme help button 21
  preview button 20, 21
header type 216
hide bar 21
HTML filters
  about 132
  product gallery, building 135-139
  reference link 139
  working with 132-134
html input 196, 197

I

if tags 58-62
image_picker input 204-206
increment tag 68, 70
input setting
  attributes 184
integer (int) 39

iteration tags
  about 71
  cycle tag 78-81
  for/else tags 71-73
  for parameters 74
  jump statements 73

J

javascript tag 259
JSON filter 172
JSON settings
  exploring 180-184
JSON template
  pages, enhancing with 243
  structure, building 244-252
  upgrading, with metafields 252-255
jump statements
  about 73
  break statement 74
  continue statement 73

L

layout directory
  about 23
  checkout.liquid 26
  gift_card.liquid 23
  password.liquid 23-25
  theme.liquid 25
layout tag 82, 83
limit parameter 75
linklist input 197, 198
Liquid
  about 32
  delimiters 33
  dynamic content 32
  static content 32

liquid input 198
liquid tag 83, 84
liquid variables 96
locales directory 29
logic operators
  working with 36-38

M

main content element 82
math filters
  about 164, 165
  product discount price 166-169
  reference link 169
max_blocks attribute 242, 243
metafields
  about 117
  app, setting up 118-122
  used, for upgrading JSON
    template 252-255
  value, rendering 123, 124
  worflow 117
metafields, elements
  key attribute 117
  namespace 117
  value 117
Metafields Guru app 119
money filters
  about 164, 165
  product discount price 166-169
  reference link 169

N

name attribute 228, 229
navigation menu
  updating 58

nesting navigation
  reference link 107
nil 40, 41
number 39
number input 186

O

offset parameter 76
or tags 62, 63

P

page input 214
pages
  enhancing, with JSON templates 243
paginate tag 87, 88
paragraph type 218, 219
password protection
  reference link 12
POST request
  /cart/add.js endpoint 266-271
  /cart/change.js endpoint 273-275
  /cart/clear.js endpoint 276, 277
  /cart/update.js endpoint 272, 273
  used, for updating cart session 265
predictive search parameters
  reference link 287
presets attribute 232, 233
preview bar 20
product accordions
  building 153-163
product input 215
product media gallery
  enhancing 140-152
product page
  creating 54, 55

Progressive JPEG 137

R

radio input 187, 188
range input 188, 189
range parameter 77
raw tag 91
render tag 88-90
reversed parameter 78
richtext input 194-196

S

section
  about 224
  static, versus dynamic sections 224-228
section directory 28
section object
  reference link 233
section schema
  class attribute 230
  name attribute 228, 229
  presets attribute 232, 233
  settings attribute 230, 232
  working with 228
section-specific tags
  exploring 256
  javascript tag 259
  stylesheet tag 256
  style tag 256-258
select input 190, 191
settings attribute 230, 232
settings_data.json file 180
settings_schema.json file 180
Shopif URL
  submitting, to browser 33

Shopify
    about 4
    learning 5-7
Shopify Admin API 265
Shopify Ajax API 264, 265
Shopify, domains
    reference link 11
Shopify, menus and links
    reference link 11
Shopify plus 25
Shopify UI Elements
    reference link 117
Shopify, webpages
    reference link 11
Sidebar action 22
Sidebar action, directories
    about 22
    config directory 28
    layout directory 23
    locales directory 29
    section directory 28
    snippet directory 28
    template directory 26-28
sidebar settings
    about 216
    header type 216
    paragraph type 218, 219
snippet directory 28
snippet input 220
social sharing images
    reference link 11
spam protection
    reference link 12
specialized input settings
    about 194
    article input 211, 212
    blog input 212, 213
    collection input 213, 214

color input 199, 200
font_picker input 206-210
html input 196, 197
image_picker input 204-206
linklist input 197, 198
liquid input 198
page input 214
product input 215
richtext input 194-196
url input 201, 202
video_url input 202-204
special objects 124
static content 32
static sections 225-227
string 38
string filters 153
stylesheet tag 256
style tag 256-258

T

template directory 26-28
textarea input 193
text input 192
t filter
    about 170-172
    reference link 172
theme
    about 12
    managing 12-19
theme editor
    organizing 216
theme structure 20
theme tags
    about 81
    comment tag 91
    echo tag 83, 84
    form tag 85-87

layout tag 82, 83
liquid tag 83, 84
paginate tag 87, 88
raw tag 91
render tag 88-90
title and meta description
  reference link 11
truncate filter
  parameters 116

U

unless tags 65
URL filters
  about 132
  product gallery, building 135-139
  reference link 139
  working with 132-134
url input 201, 202

V

variable tags
  about 66
  assign tag 66, 67
  capture tag 67, 68
  decrement tag 70, 71
  increment tag 68, 70
video_url input 202-204

W

when tags 63, 64
whitespace
  controlling 48

www.ingramcontent.com/pod-product-compliance
Lightning Source LLC
Chambersburg PA
CBHW062059050326
40690CB00016B/3152